Emerging Technologies in Space Exploration

Matthew N. O. Sadiku, Ph.D., P.E.

Regents Professor Emeritus and IEEE Life Fellow
Prairie View A&M University
Prairie View, TX 77446
Email: sadiku@ieee.org
Web: www.matthew-sadiku.com

BOOK FILMS MEDIA

Date published: January 24, 2025

Matthew N. O. Sadiku

sadiku@ieee.org
www.matthew-sadiku.com

Paperback: 979-8-9924816-1-7
eBook: 979-8-9924816-2-4
Hardcover: 979-8-9924816-3-1

BookFilmsMedia

2780 South Jones Blvd Suite 200 - 4007
Las Vegas, NV 89146 United States
+1 725 - 238 - 6534

Front cover from *NASA*

In the photo taken on Sept. 16, 1993, NASA astronauts James H. Newman (left), and Carl E. Walz evaluate procedures and gear for an upcoming Hubble Space Telescope servicing mission with the Caribbean Sea and part of the Bahama Islands chain in view. Newman and Walz's spacewalk, part of the STS-51 mission, lasted seven hours, five minutes and 28 seconds.

https://www.nasa.gov/image-article/a-caribbean-spacewalk/

DEDICATED TO MY WIFE:

JANET O. SADIKU

BRIEF TABLE OF CONTENTS

PREFACE

Space exploration has always captured our imagination, and it continues to be an essential avenue for scientific discovery and human achievement. It has always been of interest to scientists and governments across the globe as it holds the key to the origin of mankind and many marvelous wonders of the universe including the possibility of alien lives. It conjures up images of rockets blasting off to distant planets. Space exploration is more exciting and complex than ever as we strive to reach the stars. It addresses the fundamental questions about our universe and the history of our solar system. Imagine floating in zero gravity, gazing down at earth, and experiencing the awe-inspiring vastness of space. The evolution of human interest in space has been driven by the desire to explore the unknown, pushing the boundaries of knowledge with the urge to discover new worlds.

Space exploration and inhabiting other celestial bodies are important milestones in technological progress. It is one of the latest innovative and rapidly growing sectors of human intelligence. Space exploration requires cutting-edge technology that pushes the boundaries of what we thought was possible. Technology has been the key enabler of space exploration. The field of space technology is an ever-evolving and highly dynamic modern industry. Emerging space technologies are transforming societies, economies, and geopolitics. Such emerging technologies include artificial Intelligence, robotics, 3D printing, blockchain, gamification, the Internet of things, cloud computing, and big data.

This book examines these emerging technologies and their impact on space exploration. It is organized into twelve chapters that summarize emerging technologies and their applications in space exploration.

Chapter 1 - Introduction:

This chapter explores emerging technologies used in the space industry and serves as an introduction to the rest of the book. It discusses some emerging technologies in space exploration. Emerging technology is a term generally used to describe new technology. Traditional space technologies will include satellites, antennas, materials science, propulsion, rockets, and communication systems. The space industry utilizes emerging technologies, like artificial intelligence, 3D printing, 5G, Internet of things, robotics, blockchain, and cloud computing, to enhance operations in space.

Chapter 2 - Artificial Intelligence:

This chapter examines the use of artificial intelligence in space exploration and research. Artificial intelligence (AI) refers to the simulation of human intelligence in machines that are programmed to think like humans and mimic their actions. In space exploration, AI becomes critical in overcoming the challenges of long communication lags, handling massive datasets, and enabling autonomous robotic planetary exploration systems. AI's role in space exploration is transformative, processing vast data from celestial bodies and predicting risks like solar storms and space debris.

Chapter 3 - Robotics:

In this chapter, we delve into the fascinating roles of robots in space exploration. Robotics is the discipline of designing and constructing machines, called robots. It plays a crucial role in space exploration by enabling the exploration of distant planets and celestial bodies that are too hazardous or difficult for humans to reach. It extends human capabilities, reduces risks, enhances efficiency and safety in missions, and paves the way for future human exploration. Robots offer unparalleled resilience, unmatchable precision, and evolving autonomy, positioning them as indispensable tools in the future of space exploration.

Chapter 4 - Drones:

This chapter explores major applications of drones in space exploration. Drones are increasingly being used in space exploration, transportation, and other applications. They are reshaping our understanding of the universe and revolutionizing the space industry by making it more cost-effective, efficient, and accessible. Drones have the potential to revolutionize future space missions, enabling greater understanding and exploration of the universe. Using drones to explore the other planets or moons is one of the main priorities of space agencies.

Chapter 5 - Big Data:

In this chapter, we will delve into the fascinating role of big data in space exploration. Big data applies to data sets of extreme size (e. g. exabytes, zettabytes) which are beyond the capability of the commonly used software tools. It plays a crucial role in space exploration by enabling scientists to analyze vast amounts of data collected from telescopes, satellites, and space probes. The role of big data in space exploration is undeniable, revolutionizing the way we observe, analyze, and understand the universe. As we continue to push the frontiers of space exploration, there is no doubt that big data will remain a cornerstone of astronomical research.

Chapter 6 - Cloud Computing:

This chapter reviews the role cloud computing plays in the space exploration. Cloud computing is the on-demand delivery of computing resources, such as storage, processing power, and applications, via the Internet. In the space exploration, cloud computing plays a pivotal role in managing the vast amounts of data generated by satellites, telescopes, and other space-based assets. Cloud computing is crucial for space exploration because all the new missions emerging in space in the next 10 or 20 years are going to require cloud-based technology.

Chapter 7 - Internet of Things:

This chapter examines the integration of IoT in space exploration. Internet of things (IoT) refers to the idea of connecting virtually any electronic device to the Internet and connecting them through the Internet to each other. The integration of the IoT in space exploration and colonization represents a major leap forward in technological innovation, allowing for continuous data collection, automated operations, and enhanced scientific discovery. IoT enables remote monitoring in space missions by using sensors and communication networks to collect and transmit data from spacecraft and equipment continuously.

Chapter 8 - Blockchain:

In this chapter, we embark on a journey to unravel the implications of marrying blockchain technology with space concepts. Blockchain, the backbone of cryptocurrencies like Bitcoin, is a distributed ledger technology, in which blocks of data are chained together by cryptographic hashing. Blockchain can serve as one of the potential technology solutions to help remediate some of the challenges confronting space exploration. As humanity ventures further into the cosmos, blockchain may become as integral to space exploration as rockets and rovers. Blockchain technology is poised to play an increasingly important role in shaping the space industry's future.

Chapter 9 - Nanotechnology:

This chapter examines the implication of nanotechnology in space and its effects on geopolitics. Nanotechnology refers to the characterization, fabrication and manipulation of structures, devices or materials that have one or more dimensions that are smaller than 100 nanometers. It is working with things at the nanoscale. It is revolutionizing space travel and reshaping our understanding of the cosmos. It has the potential to greatly enhance the capabilities of future space missions and increase their efficiency and effectiveness. Nanotechnology may hold the key to making space flight more practical.

Chapter 10 - 3D Printing:

This chapter examines the use of 3D printing technology in space. 3D printing, also known as additive manufacturing, is a technique for making a three-dimensional object of almost any shape through additive processes in which successive layers of material are laid down under computer control to generate the precise shape of the programmed model. 3D printing in space is essential because it can manufacture intricate and customized structures locally, substantially decreasing mission expenses by launching only raw materials and essential components. In space exploration, 3D printing has simply become part of life.

Chapter 11 - Cybersecurity:

This chapter provides an introduction to cybersecurity on space vehicles. Cybersecurity refers to a set of technologies and practices designed to protect networks and information from damage or unauthorized access. The realm of space is not just a frontier for exploration but also a battleground for cybersecurity. Today, satellites and space technologies are now vulnerable to cyber attacks and espionage, such as data manipulation and signal interference. Cybersecurity in space exploration is a critical concern due to the potential for cyberattacks to disrupt space missions, endanger people, and cause international conflicts. Cybersecurity is important for space exploration because space missions are becoming more sophisticated and vulnerable to cyber threats.

Chapter 12 - Gamification:

The aim of this chapter is to describe how the gamification approach has been applied within the space sector. Gamification is a technique that incorporates game-like elements into non-game contexts to make learning more engaging. It is used to engage the public, especially students, in space missions. It can be used to train astronauts and educate the public about space. Gamification elements, such as competition, rewards, and interactive challenges, can be used to encourage exploration.

This is a comprehensive introductory text on the issues, ideas, theories, and applications of emerging technologies in space exploration. It provides an overview of each technology in simple terms so that beginners can understand it. It is a must-read book for anyone who wants to learn about modern space exploration, which have become vital in recent times.

I am grateful for the support of Dr. Annamalia Annamalai, the department head of the Department of Electrical and Computer Engineering, and Dr. Pamela Obiomon, the dean of the College of Engineering at Prairie View A&M University, Prairie View, Texas. Special thanks are due to my wife Dr. Janet Sadiku for helping in various ways.

- M. N. O. Sadiku

ABOUT THE AUTHOR

Matthew N. O. Sadiku received his B. Sc. degree in 1978 from Ahmadu Bello University, Zaria, Nigeria and his M.Sc. and Ph.D. degrees from Tennessee Technological University, Cookeville, TN in 1982 and 1984 respectively. From 1984 to 1988, he was an assistant professor at Florida Atlantic University, Boca Raton, FL, where he did graduate work in computer science. In total, he received seven college degrees. From 1988 to 2000, he was at Temple University, Philadelphia, PA, where he became a full professor. From 2000 to 2002, he was with Lucent/Avaya, Holmdel, NJ as a system engineer and with Boeing Satellite Systems, Los Angeles, CA as a senior scientist. He is presently a Regents professor emeritus of electrical and computer engineering at Prairie View A&M University, Prairie View, TX.

He is the author of over 1310 professional papers and over 140 books including "Elements of Electromagnetics" (Oxford University Press, 7th ed., 2018), "Fundamentals of Electric Circuits" (McGraw-Hill, 7th ed., 2020, with C. Alexander), "Computational Electromagnetics with MATLAB" (CRC Press, 4th ed., 2019), "Principles of Modern Communication Systems" (Cambridge University Press, 2017, with S. O. Agbo), and "Emerging Internet-based Technologies" (CRC Press, 2019). In addition to the engineering books, he has written Christian books including "Secrets of Successful Marriages," "How to Discover God's Will for Your Life," and commentaries on all the books of the New Testament Bible. Some of his books have been translated into French, Korean, Chinese (and Chinese Long Form in Taiwan), Italian, Portuguese, Spanish, German, Dutch, Polish, and Russian.

He was the recipient of the 2000 McGraw-Hill/Jacob Millman Award for outstanding contributions in the field of electrical engineering. He was also the recipient of Regents Professor award for 2012-2013 by the Texas A&M University System. He is a registered professional engineer and a life fellow of the Institute of Electrical and Electronics Engineers (IEEE) "for contributions to computational electromagnetics and engineering education." He was the IEEE Region 2 Student Activities Committee Chairman. He was an associate editor for IEEE Transactions on Education. He is also a member of Association for Computing Machinery (ACM). His current research interests are in the areas of computational electromagnetic, computer science/networks, engineering education, and marriage counseling. His works can be found in his autobiography, "My Life and Work" (Author's Tranquility Press, 2024) or his website: www.matthew-sadiku.com. He currently resides with his wife Janet in Westlake Florida. He can be reached via email at sadiku@ieee.org

TABLE OF CONTENTS

CHAPTER 1
INTRODUCTION

"God has no intention of setting a limit to the efforts of man to conquer space."

—Pope Pius XII

1.1 INTRODUCTION

Space exploration conjures up images of rockets blasting off to distant planets. It addresses the fundamental questions about our universe and the history of our solar system. Imagine floating in zero gravity, gazing down at earth, and experiencing the awe-inspiring vastness of space, as shown in Figure 1.1 [1].

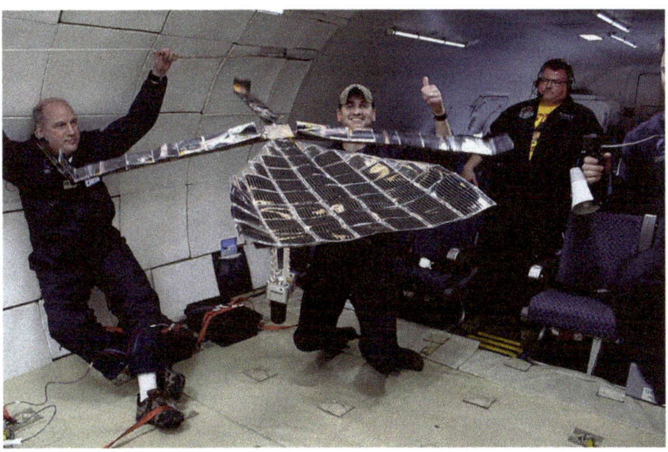

Figure 1.1 Floating in zero gravity [1].

By addressing the issues related to space exploration, humans find opportunities to advance mining, material science, and life science research. Researchers and engineers are pushing the limits of known science.

Weather forecasting, remote sensing, GPS navigation, satellite television, and long-distance communication all depend on space infrastructure. Space systems continue to become more and more integral to homeland defense, weather surveillance, communication, navigation, imaging, and remote sensing for chemicals, fires, and other disasters.

The realm of space exploration has witnessed remarkable advancements in spacecraft technology. Technology is the driving force behind the magnificent achievements and groundbreaking discoveries in space exploration. It plays a crucial role in space exploration and unlocking the mysteries of the cosmos [2].

Space exploration has always captured our imagination, and it continues to be an essential avenue for scientific discovery and human achievement. Technology has been the key enabler of space exploration. Space technology plays an important role in weather forecasting, climate monitoring, navigation, and telecommunications. Space exploration requires cutting-edge technology that pushes the boundaries of what we thought was possible. Emerging space technologies are transforming societies, economies, and geopolitics [3].

This chapter explores emerging technologies used in the space industry and serves as an introduction to the rest of the book. It begins with providing a brief history of space exploration. It covers some conventional space technologies. It describes what emerging technologies are. It mentions some emerging technologies in space exploration. It addresses the future of space exploration. It highlights the benefits and challenges of emerging technologies in space exploration. The last section concludes with comments.

1.2 BRIEF HISTORY OF SPACE EXPLORATION

The field of space technology is an ever-evolving and highly dynamic modern industry.

Throughout the 20th century, researchers were obsessed with the quest to explore the universe and to invent new space technology. More than half a century after NASA's Apollo Program enabled the first astronaut to set foot on the Moon's surface, the space industry continues to embrace game-changing technologies. In the latter half of the 20th century, rockets were developed that were powerful enough to overcome the force of gravity to reach orbital velocities, paving the way for space exploration to become a reality.

In the 1930s and 1940s, Nazi Germany saw the possibilities of using long-distance rockets as weapons. In 1957, the Soviets launched the first artificial satellite, Sputnik 1, into space. 1961, Russian Yuri Gagarin became the first human to orbit earth in Vostok 1. The first US satellite, Explorer 1, went into orbit in 1958. In 1961, Alan Shepard became the first American to fly into space. Landing a man on the Moon and returning him safely to Earth within a decade was a national goal set by President John F. Kennedy in 1961. By the early 1970s, orbiting communications and navigation satellites were in everyday use. In April 1981, the launch of the space shuttle Columbia ushered in a period of reliance on the reusable shuttle for most civilian and military space missions. The space shuttle, shown in Figure 1.2 was the first reusable spacecraft to carry people into orbit [4]. On February 1, 2003, the shuttle broke apart while reentering the Earth's atmosphere, killing all seven crew members. The final shuttle mission was completed with the landing of Atlantis on July 21, 2011, closing the 30-year space shuttle program.

Figure 1.2 The Space Shuttle [4].

One of the most prominent space missions of our times is NASA's Artemis return-to-the-moon program. The Artemis space program is widely different from the first race to the Moon of the 1950s, and 60s and 70s. This time, NASA is playing the long game and wants to establish a permanent human presence on the Moon. Modern space exploration is reaching areas once only dreamed about. Mars is focal point of modern space exploration, and manned Mars exploration is a long-term goal of the United States. The space race was then followed by an era of space cooperation, highlighted by the International Space Station. Beyond US space exploration, other nations are making technological progress, and China has become the main contender.

1.3 CONVENTIONAL SPACE TECHNOLOGIES

Technology is the driving force behind the magnificent achievements and groundbreaking discoveries in space exploration. Space technologies have the potential to guard life on Earth. Progress in space is largely reliant on technology that lowers costs while maintaining or improving upon reliability and safety. Figure

1.3 shows a deeper diving into space [5].

Figure 1.3 A deeper diving into space [5].

Traditional space technologies will include satellites, antennas, materials science, propulsion, rockets, and communication systems [1].

- *Satellites:* Satellites play a pivotal role in space exploration technologies. Satellites have come a long way since the days of the infamous "beep-boop" sounds. Today, they play a crucial role in communication, navigation, and Earth observation. Satellites play a pivotal role in space exploration technologies. Figure 1.4 shows some satellites [6].

Figure 1.4 Some satellites [6].

Small satellites have become a leading trend in SpaceTech, especially in 2025. Their miniaturization allows for cost-effective designs, and advancements in industrial technology enable mass production. Deployment of small satellites enables a wide range of missions from Earth observation to deep space exploration at a lower cost. A low-earth orbit (LEO) is relatively close to Earth's surface and is normally at an altitude of less than 1000 km but could be as low as 160 km above Earth. LEO satellites and multi-satellite constellations are increasingly in use for communication, spying, earth monitoring, and other imaging applications. For example, advanced communication systems specifically designed for LEO satellites focus on enhancing signal strength and reducing latency to ensure reliable data transmission. The latest generation of satellites is far more advanced than their predecessors, offering enhanced capabilities in communication, earth observation, and scientific research.

- *Rockets:* At the core of space exploration lies rocket technology. Early rockets provided the initial steppingstones into space, but advancements over the years have transformed space travel. More powerful engines, advanced materials, and improved fuel efficiency have allowed us to reach farther into space. Singaporean Equatorial develops a commercial sub-orbital rocket capable of delivering small payloads above the Karman line, the boundary between space and atmosphere. As we continue to push the boundaries of rocketry, innovative solutions like reusable rockets, next-generation propulsion systems, and more efficient manufacturing processes promise to make space travel more accessible and cost-effective.

- *Propulsion:* Propulsion systems are a key trend in the SpaceTech industry, offering innovative solutions for space travel. Examples include electric propulsion systems, which use electrical energy to accelerate propellant at high speeds, and green propulsion systems, utilizing environmentally friendly fuels like hydrogen and oxygen. Water-based propulsion uses water as a propellant, offering a safe and cost-effective option. There is increasing demand for advanced and sustainable propulsion solutions in space missions. Advanced propulsion is a critical aspect of making space travel feasible. Future missions demand faster travel times and energy-efficient propulsion systems. The farther into space a vehicle ventures, the more capable its propulsion systems need to be to maintain its course on the journey with precision and ensure its crew can get home. Figure 1.5 shows electric propulsion system [7].

Figure 1.5 Electric propulsion system [7].

- *Space Communication Systems:* These are a significant trend in the SpaceTech industry, focusing on advanced methods for transmitting and receiving data in space. The use of laser communication relay systems offers faster data rates and more secure communication compared to traditional radio frequency systems
- *Remote Sensing:* Space-based telescopes and remote sensing technologies have transformed our understanding of the cosmos. Telescopes like the Hubble Space Telescope and the James Webb Space Telescope have provided stunning images and critical data that deepen our knowledge of the universe's vastness and complexity.

- *Camera Phones:* In the 1990s, a team at NASA's Jet Propulsion Laboratory worked to create cameras with scientific quality that were small enough for spacecraft.
- *Space Suit:* RTX (a diverse team of explorers) has shaped the history of space exploration. The team designed the first space suit that allowed astronauts to walk on the moon, as well as the suit NASA astronauts currently use when operating outside the International Space Station. Figure 1.6 shows an astronaut with space suit [8].

Figure 1.6 An astronaut with space suit [8].

1.4 WHAT ARE EMERGING TECHNOLOGIES?

Technology may be regarded as a collection of systems designed to perform some function. It can help alleviate some of the challenges facing business today. Emerging technology is a term generally used to describe new technology. The term often refers to technologies currently developing or expected to be available within the next five to ten years. Any imminent, but not fully realized, technological innovations will have some impact on the status quo.

Emerging technologies are shaping our societies. They continue to affect the way we live, work, and interact with one another. Emerging technology (ET) lacks a consensus on what classifies them as "emergent." It is a relative term because one may see a technology as emerging and others may not see it the same way. It is a term that is often used to describe a new technology. A technology is still emerging if it is not yet a "must-have" [9]. An emerging technology is the one that holds the promise of creating a new economic engine and is trans-industrial. ET is used in different areas such as media, healthcare, business, science, education, or defense.

The characteristics of emerging technologies include the following [10]:

- *Novelty:* Emerging technologies are typically new or novel, meaning they have yet to be widely adopted or used. They often represent a significant departure from existing technologies or processes.
- *Potential for Disruption:* Emerging technologies have the potential to disrupt existing markets, industries, or ways of doing things. They may also displace existing businesses or industries.
- *Uncertainty:* Because emerging technologies are still in the early stages of development, there is often a high uncertainty surrounding their future potential and impact. It can be challenging to predict how they will evolve.
- *Rapid Change:* Emerging technologies often evolve rapidly, with new developments and innovations emerging frequently. It can make keeping up with the latest trends and advancements challenging.
- *Interdisciplinary:* Emerging technologies often involve multiple disciplines or fields of study, such as computer science, engineering, and biology. They may require collaboration across different fields and industries to develop their potential fully.

Emerging technologies are worth investigating. They are responsible for developing new products or devices. The military often looks to emerging technologies for new services or tools that will help them create a competitive business advantage.

1.5 EMERGING TECHNOLOGIES SPACE EXPLORATION

Emerging technologies usher in a new era of space exploration. The space industry utilizes emerging technologies, like artificial intelligence, 3D printing, 5G, robotics, biotechnology, and quantum computing, to enhance operations in space. We now consider these technologies in detail [11,12]:

- *Artificial Intelligence:* Artificial Intelligence (AI) has permeated various industries, transforming how businesses operate and deliver services. It has become an integral part of space missions, from enabling autonomous navigation to analyzing vast amounts of data. AI can help with many problems that accompany spaceflight. AI algorithms aid in identifying patterns, predicting space phenomena, and optimizing mission operations. The European Space Agency (ESA) notes several possible uses for AI, including helping constellations of satellites avoid collisions, analyzing large amounts of data, and enabling rovers or probes to operate autonomously when human input can be minutes or hours away because of their distance from earth. The vast amounts of data collected during space missions require sophisticated analysis. AI and machine learning have become indispensable tools for processing and interpreting the complex data sets. With AI at the helm, we can explore the furthest reaches of the universe with unparalleled precision and efficiency. While AI offers numerous benefits, it also raises ethical and social concerns. The deployment of AI must be managed with a focus on ensuring fairness, transparency, and accountability. As AI technology advances, its applications will become more widespread, making it an integral part of everyday life. The end goal of AI is to translate and fully understand human languages.
- *Robotics:* Robots are the unsung heroes of space exploration. Before sending humans into the cosmos, robotic missions

paved the way for exploration. Unmanned missions allow us to take risks, explore hazardous environments, and gather vital data without jeopardizing human lives. Future space technologies will rely heavily on robotics and AI to assist human teams of astronauts. Robotic mining equipment designed for extreme space environments is capable of autonomously drilling and extracting resources. With each advancement, the line between human and robot exploration blurs, and the collective effort to uncover the secrets of the universe becomes clearer. Figure 1.7 shows NASA space robot Spirit [13].

Figure 1.7 NASA space robot Spirit [13].

- *Drones:* These machines (flying robots) go where humans cannot, gathering data and paving the way for future missions. Drones can be used to collect samples from planets and moons. A drone can fly to a specific location, gather samples, and return to the main spacecraft, all while operating autonomously. This could be a game-changer for missions aiming to study the composition of extraterrestrial bodies. Figure 1.8 shows a typical drone [5].

Figure 1.8 A typical drone [5].

- *Big Data:* Big data analysis of planetary data collected from space probes, telescopes, and simulations allows researchers to identify patterns and trends that reveal the underlying mechanisms of planetary formation. Big data analytics are being employed to manage and process the enormous datasets collected by satellites, facilitating efficient data storage, retrieval, and utilization for various space applications.

- *Blockchain:* Cryptocurrencies are perhaps the most well-known application of blockchain technology. They are digital or virtual currencies that use cryptography for security, making them difficult to counterfeit. By providing a transparent and immutable ledger, blockchain allows for the secure and efficient tracking of products from manufacturer to consumer. This transparency helps in reducing fraud, errors, and inefficiencies. Smart contracts are self-executing contracts with the terms of the agreement directly written into code. They run on blockchain technology, allowing them to be secure and immutable. Researchers are also leveraging blockchain technology to ensure secure and tamper-proof data transmission, enhancing the reliability of communication between satellites and ground stations.

- *3D Printing:* 3D printing (also called additive manufacturing) is a new technology that has changed lots of industries. Instead of carving something out of a big block of material,

3D printing builds things layer by layer from a digital design. Manufacturing launch vehicles and satellites require precise engineering and can be a lengthy, expensive process. The 3D printed parts were used in the March 2023 launch of the Terran 1 rocket, and this additive manufacturing approach to fabrication is likely to be used in future space explorations. NASA has made a 3D print rocket engine components using a copper-based alloy. 3D printing holds significant promise as a technique that can be harnessed to prepare tailor-made medicines and even food to address specific remedies or nutritional element substitution for the well-being of space travelers. Figure 1.9 shows 3D printing on the Moon [14].

Figure 1.9 3D printing on the Moon [14].

- *Nanotechnology:* The heavier the payload, the more energy it takes to send a rocket to space. Hence the appeal of nanotechnology, a branch of engineering that aims to dramatically shrink the size of electronic components like microchips and sensors. Lower costs are not the only benefit of working with nanotechnology, as manufacturing at ever-smaller scales can pack more processing power onto a chip or sensor. Nanotechnology is all about working with super tiny stuff—atoms and molecules. Because things act differently when they are that small, it opens up a whole world of possibilities for making stuff with unique properties we cannot get at bigger sizes.

1.6 FUTURE OF SPACE EXPLORATION

The future of space exploration is here. As we gaze up at the night sky and ponder the vastness of the universe, it is hard not to get excited about the state of the space industry. The era of space exploration is no longer solely the domain of governments. Public-private partnerships have become a driving force for innovation in the industry. Gone are the days when space exploration was limited to government-funded missions. With their ambitious goals of colonizing Mars and making space tourism a reality, the industry is experiencing a renaissance like never before. Space exploration holds immense significance for the future of humanity. Whether it is Elon Musk sending a Tesla into orbit or Jeff Bezos planning to establish a lunar outpost, their larger-than-life personalities have breathed new life into the field [15]. Figure 1.10 displays the next generation space technologies [16].

Figure 1.10 Next generation space technologies [16].

Space exploration is an enabling sector that transcends the boundaries of traditional industries. This is an exciting moment for space enthusiasts. NASA's initiative to send humans back to the moon is well underway and space tourism has finally gotten off the ground. To continue our journey into the cosmos and unlock the secrets of the universe, we must embrace the spirit of innovation. Companies and researchers are constantly pushing the bounds of space technology.

1.7 BENEFITS

Space technologies are expected to become as integral to daily life. They have a much more immediate and profound impact on our daily lives than many of us imagine, including weather forecasting, climate monitoring, navigation and telecommunications. There are so many space applications that really affect our everyday lives. Night-time imagery can help monitor unregulated fishing, track sea ice movements, and help reduce light pollution. Other benefits include the following [15]:

- *Automation:* Process automation and robotics are subfields of AI technology that involve programming computer applications to replace many human activities for more productivity. Automation plays a crucial role in the space industry, particularly for long-term exploration and missions. The impact of AI on employment, with the automation of certain jobs, also presents a social challenge.
- *Privatization of Space:* From science to manufacturing, exploration, and even space travel, private companies are stepping up and alleviating governments from their monopolistic role while bringing in new resources and capital influx. The privatization of space has started recently, and we will see new and established players engaging in healthy competition and pushing space missions and technology further than ever before.
- *Space Travel:* This is a distant dream for most, but space technology is already playing a crucial role in our daily lives.
- *Space Tourism:* Space tourism is no longer a pipe dream. It is rapidly becoming a reality. Companies like Virgin Galactic and Blue Origin are pioneering this industry, offering suborbital flights to private individuals. Thanks to the efforts of private companies, space tourism is on the cusp of becoming a reality. Commercial space travel companies are the driving force behind the future of space tourism. As with any new industry, space tourism comes with its fair share of challenges and considerations. Figure 1. 11 shows space tourism [7].

Figure 1.11 Space tourism [7].

- *Global Challenges:* Space technologies can also help to address global challenges, including disaster response and climate monitoring.

1.8 CHALLENGES

Space exploration and utilization have historically been fraught with technical challenges. The reality of space travel is far from accessible for most people and remains largely a "billionaire's dream. " While consumers may remain largely in the dark on the role played by space technology, industry leaders recognize its impact. Other challenges include the following [15]:

- *Safety:* Safety tops the list of challenges, as strapping yourself to a rocket and hurtling into space is no walk in the park. There are also ethical concerns about the environmental impact of space tourism and the potential exploitation of celestial bodies.

- *Space Debris:* Space is increasingly cluttered with debris. The growing presence of space debris necessitates innovative solutions to prevent potential collisions and ensure the long-term sustainability of space activities.

- *Human Colonization:* Mars has proven to be a treasure trove of scientific discoveries. But colonizing Mars is no picnic. With its thin atmosphere, extreme temperatures, and lack of readily available resources, it is a hostile environment for humans. However, plans are already in motion to overcome these challenges. In space, there are no borders or rivalries, only a shared fascination with the unknown.
- *Climate Change:* This is not just a terrestrial issue; its effects ripple through the cosmos. Understanding the impacts of climate change on space exploration is crucial for mitigating risks and planning future missions. Extreme weather events have implications for launching rockets, satellite communication, and the overall sustainability of space activities. It is a reminder that our actions on earth have consequences that reach far beyond our atmosphere.
- *Heat:* Going to the Moon is no easy task. The farther a spacecraft travels in space, the more heat it will generate as it returns to Earth. Getting back safely requires technologies that can help a spacecraft endure speeds 30 times the speed of sound and heat twice as hot as molten lava or half as hot as the sun.
- *Sustainability:* Space is not just a vast expanse of emptiness. Sustainability concepts that have never been applied to the space sector are expected to become the norm. Sustainable practices in space exploration ensure that we leave as small a footprint as possible. As responsible space explorers, we must adapt and mitigate the impacts of climate change on our cosmic endeavors. This means embracing sustainable practices, reducing carbon emissions, and investing in greener technologies.
- *Regulation:* In the vastness of space, rules and regulations are necessary to ensure safety, fairness, and responsible behavior. Regulating space activities is a significant challenge due to the international nature of the industry and the absence

of clear legal frameworks. International organizations, such as the United Nations Office for Outer Space Affairs, work tirelessly to establish norms and guidelines for space exploration. They are to ensure that we explore the cosmos responsibly and ethically.

- *Collaboration:* This goes hand in hand with space exploration, and international agreements and legal frameworks are the vehicles that drive this collaboration forward. Treaties, conventions, and guidelines ensure that all nations have an equal opportunity to explore space while respecting the rights of others. These agreements also help regulate activities such as launching rockets, operating satellites, and protecting celestial bodies. By coming together and playing by the same rules, we can create a harmonious celestial neighborhood for all.

1.9 CONCLUSION

Space exploration addresses the fundamental questions about our universe and the history of our solar system. Emerging technologies are intersecting and interacting in a host of ways, with important implications for policy. The future promises greater integration of these technologies into everyday life.

Modern spacecraft design has evolved significantly. Today's spacecraft are equipped with advanced materials for durability, innovative propulsion systems for faster travel, and sophisticated instruments for detailed exploration. New space technology aims to lower costs and expand access to orbit and beyond. Future space technologies will include advancements in manufacturing, materials science, propulsion, and artificial intelligence to expand our capabilities in earth orbit and on interplanetary missions. The future of space exploration is bright with potential. From plans to return humans to the Moon to ambitious missions to Mars and beyond, the possibilities are limitless. More information on emerging space technologies is available from the books in [17,18] and a related journal: *Journal on Emerging Technologies.*

REFERENCE

[1] "Small spacecraft technology,"

https://www.nasa.gov/smallspacecraft/

[2] "The indispensable role of technology in space exploration," August 2023,

https://nstxl.org/the-indispensable-role-of-technology-in-space-exploration/#:~:text=August%2014%2C%20 2023&text=From%20the%20first%20successful%20 mission,the%20mysteries%20of%20the%20cosmos.

[3] M. N. O. Sadiku, P. A. Adekunte, and J. O. Sadiku, "Emerging technologies in space exploration," *International Journal of Trend in Scientific Research and Development*, vol. 8, no. 6, November-December 2024, pp. 474-482.

[4] "A brief history of space exploration,"

https://aerospace.org/article/brief-history-space-exploration

[5] H. Lin and S. D'Amico, "A deeper dive into space," June 2024,

https://setr.stanford.edu/news/deeper-dive-space

[6] J. Wood, "The technologies ushering in a new era of space exploration," August 2024,

https://spectra.mhi.com/the-technologies-ushering-in-a-new-era-of-space-exploration

[7] "Emerging technologies in aerospace: Shaping the future of flight,"

https://wiia.ac.in/emerging-technologies-in-aerospace-shaping-the-future-of-flight.php

[8] "Intelligent space technologies,"

https://www.rtx.com/who-we-are/we-are-rtx/space

[9] M. Halaweh, "Emerging technology: What is it?" *Journal*

of Technology Management & Innovation, vol. 8, no. 3, 2013, pp. 108-115.

[10] N. Duggal, "Top 18 new technology trends for 2023," July 2023,

https://www.simplilearn.com/top-technology-trends-and-jobs-article

[11] "5 Next-generation space technologies being tested today," March 2022,

https://www.ni.com/en/solutions/aerospace-defense/space-launch-exploration/next-generation-of-space-technology-is-being-tested-today.html?srsltid=AfmBOopKD9TJl-4JXFVRdqUgIBkLyuWTOJAOzruFFfFqrLlkmhooPSeq

[12] "20 Emerging technologies that will change our world," January 2024,

https://blog.emb.global/emerging-technologies-that-will-change-our-world/

[13] "Spirit/Opportunity," December 2009,

https://www.nasa.gov/audience/foreducators/robotics/imagegallery/r_spirit.jpg.html

[14] A. Townsend, "Artificial intelligence for space exploration," October 2019,

https://medium.com/@apltownsend/artificial-intelligence-for-space-exploration-703d2838e37c

[15] E. C. Nwankwo, "The future of space exploration: Exploring the expanding space industry," November 2023,

https://ericnwankwo.medium.com/the-future-of-space-exploration-exploring-the-expanding-space-industry-07570013e04a

[16] "5 Next-generation space technologies being tested today,"

https://www.ni.com/en/solutions/aerospace-defense/space-launch-exploration/next-generation-of-space-technology-is-being-tested-today.html?srsltid=AfmBOooloyDXA4IxrLXnrRMwzmg5_8sqRuh9ds1Ql-u2EwZPTAC16qKR

[17] M. Razani, *Commercial Space Technologies and Applications: Communication, Remote Sensing, GPS, and Meteorological Satellites*. Boca Raton, FL: CRC Press, Second Edition, 2018.

[18] S. Tkatchova, *Space-based Technologies and Commercialized Development: Economic Implications and Benefits*. Engineering Science Reference, 2011.

CHAPTER 2
ARTIFICIAL INTELLIGENCE IN SPACE EXPLORATION

"If the government regulates against use of drones or stem cells or artificial intelligence, all that means is that the work and the research leave the borders of that country and go someplace else."

—Peter Diamandis

2.1 INTRODUCTION

Space exploration, a realm of boundless potential, has captured human imagination for generations. Space exploration has always been a primary goal for humanity, pushing the boundaries of our understanding and opening up new frontiers. In recent years, data science has emerged as a pivotal tool, transforming the way we approach space technology. Data analytics is important in contemporary space missions, encompassing data collection, processing, analysis, and interpretation.

Space exploration generates vast amounts of data — astronomy, earth science, space weather — spacecraft now generate terabytes of information weekly. The large datasets from satellites and space vehicles are often too massive to be processed manually by humans. This is where artificial intelligence comes in; deep learning algorithms, data analytics, and image processing techniques can quickly analyze this data. The importance of artificial intelligence in space exploration is significant, just like in other aspects of life. Artificial intelligence, which makes our lives easier in every field, also provides great support while exploring space. It has emerged as a critical feature in allowing humans to better understand what we have found out in the cosmos. With the beginning of the 21st century, AI has made a steady progress in Mars exploration. The ambitions for humans to reach Mars and travel beyond into our Solar System will only be made possible with the development and use of AI for Space Exploration. AI empowers our spacecraft, telescopes, and instruments to unveil cosmic secrets. As AI advances, it expands our cosmic understanding and inspires innovation on earth.

Artificial intelligence (AI) refers to the simulation of human intelligence in machines that are programmed to think like humans and mimic their actions. It has been making waves in recent years, enabling us to solve problems faster than traditional computing could ever allow. AI is transforming numerous industries, and space exploration is no exception. In space exploration, AI becomes critical in overcoming the challenges of long communication lags, handling massive datasets, and enabling autonomous robotic planetary exploration systems. AI has revolutionized space exploration, transforming the way we navigate and understand the cosmos. AI silently pioneers uncharted territories in space exploration [1].

This chapter examines the use of artificial intelligence in space exploration and research. It begins with explaining what AI is all about. It describes space exploration. It presents some applications of AI in space exploration. It highlights the benefits and challenges of AI in space exploration. The last section concludes with comments.

2.2 WHAT IS ARTIFICIAL INTELLIGENCE?

The term "artificial intelligence" (AI) is an umbrella term John McCarthy, a computer scientist, coined in 1955 and defined as "the science and engineering of intelligent machines." It refers to the ability of a computer system to perform human tasks (such as thinking and learning) that usually can only be accomplished using human intelligence [2]. Typically, AI systems demonstrate at least some of the following human behaviors: planning, learning, reasoning, problem solving, knowledge representation, perception, speech recognition, decision-making, language translation, motion, manipulation, intelligence, and creativity.

The 10 U. S. Code § 2358 define artificial intelligence as [3]:

1. "Any artificial system that performs tasks under varying and unpredictable circumstances without significant human oversight, or that can learn from experience and improve performance when exposed to data sets.
2. An artificial system developed in computer software, physical hardware, or other context that solves tasks requiring human-like perception, cognition, planning, learning, communication, or physical action.
3. An artificial system designed to think or act like a human, including cognitive architectures and neural networks.
4. A set of techniques, including machine learning, that is designed to approximate a cognitive task.
5. An artificial system designed to act rationally, including an intelligent software agent or embodied robot that achieves goals using perception, planning, reasoning, learning, communicating, decision making, and acting."

AI provides tools creating intelligent machines which can behave like humans, think like humans, and make decisions like humans. The main goals of artificial intelligence are [4]:

1. Replicate human intelligence
2. Solve knowledge-intensive tasks
3. Make an intelligent connection of perception and action
4. Build a machine which can perform tasks that requires human intelligence
5. Create some system which can exhibit intelligent behavior, learn new things by itself, demonstrate, explain, and can advise to its user.

AI is not a single technology but a range of computational models and algorithms. The concept of AI is an umbrella term that encompasses many different technologies. AI is not a single technology but a collection of techniques that enables computer systems to perform tasks that would otherwise require human intelligence. The major disciplines in AI include [5]:

- *Expert systems*
- *Fuzzy logic*
- *Neural networks*
- *Machine learning (ML)*
- *Deep learning*
- *Natural Language Processors (NLP)*
- *Robots*

These computer-based tools or technologies have been used to achieve AI's goals. Each AI tool has its own advantages. Using a combination of these models, rather than a single model, is recommended. Figure 2.1 shows a typical expert system, while Figure 2.2 illustrates the AI tools. These tools are gaining momentum across every industry. Analytics can be considered a core AI capability.

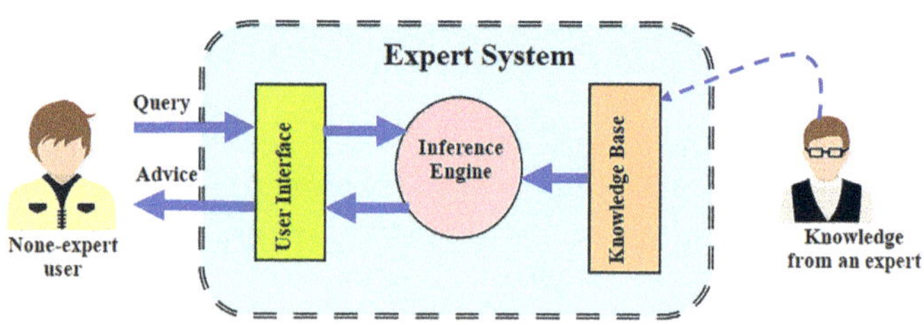

Figure 2.1 A typical expert system.

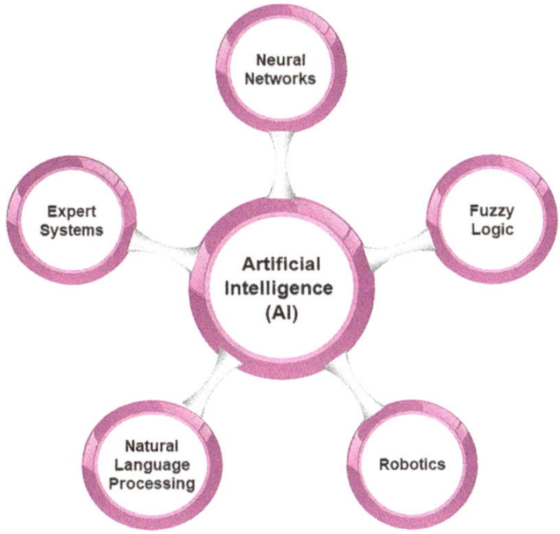

Figure 2.2 AI tools.

2.3 SPACE EXPLORATION

Space exploration has always been of interest to scientists and governments across the globe as it holds the key to the origin of mankind and many marvelous wonders of the universe including the possibility of alien lives. Scientists and engineers have explored roughly only 4% of the visible universe, which represents the parts of space that we can see using telescopes. The visible universe is made up of planets, stars, galaxies, and other astronomical objects.

Space exploration began within human minds, long before we actually went into space, we did so in our imagination. Space is an expression of humanity's endless curiosity and exploration. Every day, various space vehicles and observation systems are used to gather information about new planets, stars, and galaxies. The commercialization of space is entering an exciting new phase with AI technology integration.

One of the main aims of NASA at the moment is to increase the exploration of space. NASA extensively utilizes artificial intelligence and machine learning in space exploration, primarily for autonomous navigation of spacecraft, analyzing large datasets from missions, identifying patterns in planetary surfaces, and making real-time decisions on rovers (robots). NASA has been safely using artificial intelligence for decades to plan and schedule missions for planetary rovers. NASA has previously used AI in human spaceflight, scientific analysis, and autonomous systems. Currently, NASA is in the development phase of an assistant that would provide help with the process of engineering future missions. NASA has also previously developed an AI assistant named Daphne. This assistant focuses on providing instant feedback on questions and queries related to the earth. NASA's mission of exploration requires leveraging new ways to utilize and learn from the unprecedented amount of data that space-based observation platforms generate [6].

2.4 ARTIFICIAL INTELLIGENCE IN SPACE EXPLORATION

The exponential rise in space data captured from satellites, telescopes, and interplanetary probes requires AI's analytical capabilities. Modern space instruments generate terabytes of data daily - far more than scientists can examine manually. Artificial Intelligence represents a pivotal breakthrough in scientific advancement, redefining the boundaries of research and exploration. AI has the potential to revolutionize space exploration in many ways. Figure 2.3 illustrates AI in space exploration [7]. Some AI applications in space exploration are shown in Figure 2.4 [8].

Figure 2.3 AI in space exploration [7].

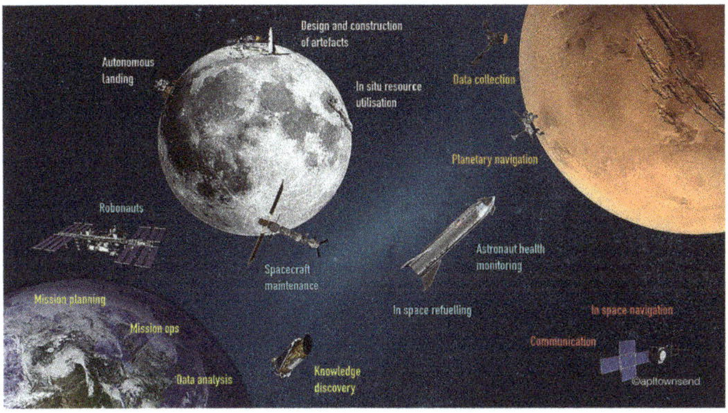

Figure 2.4 Some AI applications in space exploration [8].

Here are some of the ways that AI is being used in space exploration today [9-11]:

- *Astronaut Assistants:* Scientists and researchers are working towards creating intelligent assistants to help astronauts in their mission to Moon, Mars, and beyond. AI has become a focal point for scientists developing intelligent assistants to support astronauts in real space missions. These AI-supported assistants aim to enhance the safety and efficiency of space missions by addressing various challenges that astronauts may encounter. They are designed to understand and predict the requirements of the crew and comprehend astronauts' emotions and their mental health and take necessary actions in the case of an emergency. AI also assists with entry, descent, and landing (EDL) - the riskiest phase facing probes sent to Mars. They can detect hazardous situations such as increased carbon dioxide levels in spacecraft and provide astronauts with alerts. A typical AI-based assistant is shown in Figure 2.5 [12].

Figure 2.5 A typical AI-based assistant [12].

- *Astronaut Health Monitoring:* The mental and physical toll during missions creates a need for enhanced astronaut medical care. AI shows promise for improving future crew support systems. By integrating multi-modal data streams - from sensors tracking heart rate and skin temperature to recording exercise and sleep patterns - predictive health analytics powered by AI can enable customized interventions tailored to each astronaut. AI-based medical care can also be applied to space exploration. As humans ready themselves to leave LEO, new medical care systems that make astronauts more autonomous for delivering their own care, are highly important. Monitoring and assessing the health of astronauts on long-duration missions is a crucial challenge for space actors currently. Health is a huge barrier to overcome on deep space exploration missions to Mars. AI can be used to identify and classify health data in real-time to support the efforts of on the ground medical officers.

- *Space Debris:* The increasing amount of space debris poses a serious threat to future space missions. Space debris or space junk is any piece of machinery or debris left by humans in space. It can refer to big objects such as dead satellites that have failed or been left in the orbit at the end of their mission. The problem with space debris has reached a critical point as scientists and researchers continue to send satellites into space, which is never brought back. The major concern with these space debris is that can cause space accidents. Figure 2.6 shows the image released in 2013 by NASA which showed the amount of space debris we had back in 2013 [9].

Figure 2.6 The amount of space debris we had back in 2013 [9].

- *Data Analysis:* Space exploration generates immense amounts of data across various disciplines, including astronomy, earth science, and space weather. The capability of AI to process and analyze data efficiently onboard spacecraft represents a major advancement in space research. AI can be used to analyze large amounts of data from space missions, such as images and sensor readings. This can help scientists to better understand the planets and moons that we are exploring, and to identify potential hazards. This may involve sorting through massive amounts of scientific data to find interesting or unusual discoveries that require further investigation. Since space exploration involves huge distances, it is crucial to be able to process and analyze data quickly on the spot. AI's ability to do this helps with making timely decisions, especially during critical parts of a mission. AI's ability to process vast volumes of data enables split-second threat detection and emergency responses, essential for addressing hurdles in space exploration and empowering autonomous robotic planetary systems.

- *Mission Planning:* AI can be used to plan space missions, such as determining the best trajectory for a spacecraft or the best way to land on a planet. This can help to reduce the risk of accidents and to ensure that missions are successful. AI can help us to explore new worlds, to learn more about the universe, and to ensure that our missions are safe and successful. For example, NASA is using AI to help plan future missions to Mars. Figure 2.7 shows the space exploration mission architecture [8].

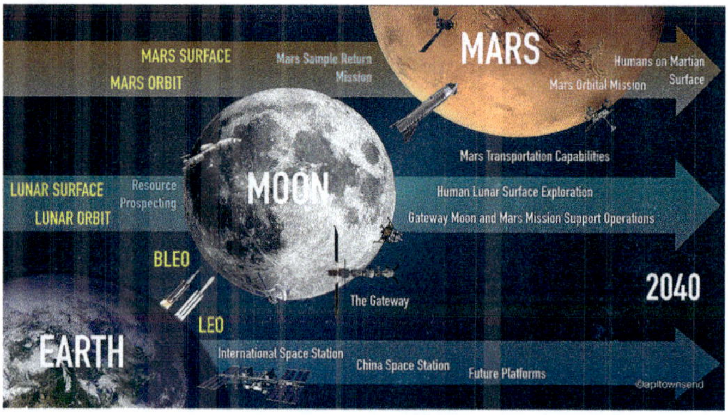

Figure 2.7 Space exploration mission architecture [8].

- *Robotics:* Robotic systems have proven to be well suited for repeated tasks in harsh and hazardous environments. Robots are used for a variety of different space applications, and their function can be divided into four main categories: Rover, Arm, Sampler, or Drill. AI can be used to control robots that are exploring space, allowing them to perform tasks that would be too dangerous or difficult for humans, such as exploring the surface of Mars. For example, the European Space Agency (ESA) is using AI to control a robot that is exploring the surface of Mars. The robot, called ExoMars, is equipped with a camera that is controlled by AI. AI-powered robotic rovers and landers such as NASA AI projects Curiosity and Perseverance navigate celestial terrains independently, gathering samples, running experiments, and snapping detailed pictures. Robots have been used for space missions for decades due to their ability to work in harsh and hazardous environments that would be unsafe for humans. Curiosity rover (robot), such as shown in Figure 2.8, can be sent to explore distant worlds such as Jupiter, where the lag between earth and spacecraft communications is too great to allow for a manual human operated landing [8]. The rovers are currently roaming the surface of Mars and making decisions without specific commands from the mission control.

Figure 2.8 Curiosity rover (robot) designed by NASA [8].

- *Satellite Navigation:* Another area where artificial intelligence comes in handy is optimizing satellite operations, paying close attention to collision avoidance and telemetry analysis. AI is also being used to help identify rogue drones in sensitive airspace. AI's role extends to optimizing satellite operations, focusing on collision avoidance and telemetry analysis. AI for data processing can also be used for the satellites themselves. AI offers huge potential to augment in-space navigation, primarily probes exploring the extremities of our solar system. Satellites observe various regions of the planet and then submit the data to ground offices.

- *Earth Observation:* This is one area where AI is already being used more extensively. Earth observation satellites generate tremendous amounts of data. This is received by ground stations in chunks over a large period of time, and has to be pieced together before it can be analyzed. Figure 2.9 shows a typical earth observation [13].

Figure 2.9 A typical earth observation [13].

2.5 BENEFITS

AI's strong points lie in its ability to handle large amounts of data quickly, identify complex patterns easily, and deal with repetitive tasks without getting tired. AI in space exploration plays a major role in enhancing efficiency, advancing scientific discovery, and ensuring the success of space missions. AI can help manage communication between space and earth by selecting the best frequencies and power levels for sending signals. AI can help establish a conversation with scientists by applying deep learning to facial recognition and speech recognition. Other benefits of AI in space exploration include [13,14]:

- *Automation:* Autonomous navigation is a key application of AI that would help us navigate around earth and other planets. Spacecraft equipped with AI capabilities can navigate space autonomously. AI automation assists with classifying and processing streams of images, sensor readings, and spectral data. AI holds immense promise for automating spectral data analysis from future missions to places like Saturn's moon Enceladus. Satellites orbiting earth also require more autonomy, as they need to make more frequent collision avoidance maneuvers to evade increasing amounts of space debris.
- *Decision Making:* AI's ability to make decisions comes into play during critical mission phases. AI can help make decisions in space missions, such as when to abort a mission

or when to change course. Onboard AI can evaluate potential actions based on mission goals and choose the best response. This autonomous decision-making reduces the need for control from earth, which is crucial for deep space exploration where signal delays can be significant.

- *Anomaly Detection:* This is another area where AI proves extremely useful, helping with identifying unusual readings from instruments. By providing early warnings for potential issues, AI enhances the safety and reliability of missions.
- *Smart Space Driving:* AI helps robots and rovers drive themselves on other planets. It is like having a really smart driver who can avoid rocks and find the best paths.
- *Assist Humans:* Space exploration is a large-scale and complex process, and human skills can sometimes be insufficient in areas such as data analysis, navigation, mission planning, and risk management required by this process. It is precisely at this point that artificial intelligence comes into play and forms a powerful alliance to push the boundaries of human exploration in space. On space stations, AI-powered robots help astronauts with daily tasks. They are like helpful assistants who never get tired. The teamwork between humans and machines, with fancy technology, has been super important for making discoveries and big progress happen.
- *Supporting Astronaut Health:* The mental and physical toll during multi-year missions creates a need for enhanced astronaut medical care. AI shows promise for improving future crew support systems. By integrating multi-modal data streams - from sensors tracking heart rate and skin temperature to recording exercise and sleep patterns - predictive health analytics powered by AI can enable customized interventions tailored to each astronaut.
- *Saving Money:* Space trips are super expensive. But AI can help cut costs by doing jobs that would need lots of people or fancy equipment.

Some of these benefits of AI in space exploration are depicted in Figure 2.10 [7].

Figure 2.10 Some of these benefits of AI in space exploration [7].

2.6 CHALLENGES

Space exploration involves numerous risks and challenges, such as long-duration missions, harsh environments, and communication delays. Some promise that AI will transform society for the better, others claim it is dangerous. How are we ensuring that it is used effectively and safely in space? One obstacle is making AI algorithms compatible with harsh space environments. Other challenges of AI in space exploration include [13]:

- *Unpredictable Space Environments:* The extreme conditions of outer space can create unique situations that pre-programmed AI may not have been trained to handle. This could potentially lead to malfunctions or failures in the system.
- *Dependence Risks:* Relying too heavily on autonomous systems could result in a decline of human expertise over time, as skills weaken without regular practice.
- *Safety:* As AI continues to advance, it is important to constantly improve its capabilities alongside ensuring the safety and innovation of space exploration. Due to

microgravity and special radiation environment, health issues may arise and could eventually peak in the inability to manage the spacecraft. AI contributes to the safety and well-being of astronauts during interstellar travel.

- *Sustainability:* If humanity wants to get the most out of AI-driven technologies, they need to be developed and used in an ethical and sustainable way.
- *Transparency:* Space agencies and tech companies should document and communicate the decision-making processes of AI systems. This transparency is vital not only for maintaining public trust but also for facilitating collaboration between different entities in the space sector.
- *Accountability:* There should be clear guidelines on liability and accountability when it comes to AI-driven decisions in space missions. Identifying responsible parties for AI actions can be complex, but it is crucial for addressing any unintended consequences or malfunctions.
- *Fairness:* The design of AI technologies must prevent biases that could lead to unfair outcomes. Although machine learning algorithms are as objective as the data they are trained on, ensuring fairness requires constant vigilance against skewed datasets or prejudiced assumptions.
- *Reliability:* AI is unproven in many aspects and lacks sufficient testing and proofing to be reliable. Interest in using AI in space exploration has been fueled by its increased reliability on missions and its potential cost savings. Relying on AI for critical activities during missions creates single points of failure within the system itself due to dependency. Redundancy measures must be put in place alongside fail-safe mechanisms so that these dangers can be mitigated while still ensuring the mission's success, even if something goes wrong with AI systems. AI models used for space exploration require a 100% robustness guarantee to satisfy project executives-for all remote applications the software simply cannot fail.

- *Risks of Cybersecurity:* Space-based AI systems are also prone to cyber threats and data privacy breaches. The risk of such incidents is greater with the increasing integration of artificial intelligence into space systems. The complex nature of AI algorithms and the interconnectedness of various networks within space create vulnerabilities that hackers can exploit, leading to the compromise of mission-critical operations.
- *Space Weather Forecasting:* Space weather, such as solar flares and cosmic rays, poses significant challenges to space exploration. AI algorithms can analyze space weather data and provide accurate forecasts, allowing spacecraft operators to prepare for conditions that might affect performance and astronauts can be alerted to take preventive measures.

2.7 CONCLUSION

AI's role in space exploration is transformative, processing vast data from celestial bodies and predicting risks like solar storms and space debris. The story of space exploration is being continuously rewritten by the influence of AI. AI is regarded as a game changer in the quest for knowledge and understanding of space. Concerted efforts persist in employing AI's capabilities to benefit space exploration. Space colonization and terraforming may have a bright future thanks to AI-driven innovations.

In the future, space agencies and private companies will continue to explore Mars and other planets in the solar system. AI offers huge potential to augment current and future space exploration. The future of space exploration depends on human beings working closely together with various AI capabilities. The collaboration between human ingenuity and AI capabilities promises to drive future discoveries and push the boundaries. AI is ready to fuel a new dawn of space exploration. More information on artificial intelligence in the space industry is available from the books in [16-27] and the following related journals:

- *Energy and AI*
- *The AI Journal*
- *Progress in Aerospace Sciences*

REFERENCE

[1] M. N. O. Sadiku, P. A. Adekunte, and J. O. Sadiku, "Artificial Intelligence in Space Exploration," *International Journal of Trend in Scientific Research and Development*, vol. 8, no. 6, November-December 2024, pp. 464-473.

[2] M. N. O. Sadiku, "Artificial intelligence," IEEE Potentials, May 1989, pp. 35-39.

[3] "Artificial intelligence (AI),"

https://www.law.cornell.edu/wex/artificial_intelligence_(ai)

[4] "Artificial intelligence tutorial,"

https://www.javatpoint.com/artificial-intelligence-tutorial

[5] D. Quinby, "Artificial intelligence and the future of travel," May 2017,

https://www.phocuswright.com/Travel-Research/Research-Updates/2017/Artificial-Intelligence-and-the-Future-of-Travel

[6] K. Sandanshi, "NASA using AI and machine Learning for space exploration," October 2020,

https://medium.com/@sandanshikunti/nasa-using-ai-and-machine-learning-for-space-exploration-e61af699307a

[7] A. Rathod, "AI in space exploration: Use cases, impact & future," July 2024,

https://aiperceiver.com/ai-in-space-exploration-use-cases-impact-future/

[8] A. Townsend, "Artificial intelligence for space exploration," October 2019,

https://medium.com/@apltownsend/artificial-intelligence-for-space-exploration-703d2838e37c

[9] "Artificial intelligence in space exploration," January 2021,

https://www.analyticsvidhya.com/blog/2021/01/artificial-intelligence-in-space-exploration/

[10] G. Chen, "AI and space exploration," December 2023,

https://drupaldemo.harvardsites.harvard.edu/news/2023/12/ai-and-space-exploration

[11] N. Sahota, "AI in space exploration: Rocketing towards new horizons," April 2024,

https://www.neilsahota.com/ai-in-space-exploration-rocketing-towards-new-horizons/#:~:text=AI's%20Indispensable%20Contribution,-In%20contemporary%20space&text=AI%2Dpowered%20autonomous%20spacecraft%20navigate,and%20exploration%20of%20distant%20galaxies.

[12] V. Kulshrestha, "AI in space exploration: Expanding the cosmos with artificial intelligence," October 2023,

https://www.linkedin.com/pulse/ai-space-exploration-expanding-cosmos-artificial-vaibhav-kulshrestha#:~:text=%F0%9F%8C%A0%20Artificial%20intelligence%20silently%20pioneers,and%20inspires%20innovation%20on%20Earth.

[13] O. Ali., "How is AI being used in space exploration?"

https://www.azoquantum.com/Article.aspx?ArticleID=474

[14] J. Canda, "The future of AI in space exploration and research," April 2024,

https://medium.com/@jam.canda/the-future-of-ai-in-space-exploration-and-research-ff2b20f2dc9c

[15] "Artificial intelligence in space,"

https://www.esa.int/Enabling_Support/Preparing_for_the_Future/Discovery_and_Preparation/Artificial_intelligence_in_space

[16] M. N. O. Sadiku, S. M. Musa, and S. R. Nelatury, *Applications of Artificial Intelligence*. Sherida, NY: Gotham Books, 2022.

[17] HAL, *Space in AI (Artificial Intelligence): Part I: Earth*

Orbit. Kindle Edition, 2024.

[18] A. A. Desrochers (ed.), *Intelligent Robotic Systems for Space Exploration (The Springer International Series in Engineering and Computer Science, 168)*. Springer, 2nd edition, 1992.

[19] M. Madi and O. Sokolova, *Artificial Intelligence for Space: AI4SPACE: Trends, Applications, and Perspectives*. Boca Raton, FL: CRC Press, 2023.

[20] C. Ieracitano et al. (eds.), *The Use of Artificial Intelligence for Space Applications Workshop at the 2022 International Conference on Applied Intelligence and Informatics. Springer, 2023.*

[22] J. C. Lok, *Prediction How Artificial Intelligence Impacts to Space Development?* Amazon Digital Services LLC, 2019.

[23] J. C. Lok, *Ethic To Artificial Intelligence Space Development?* Independently Published, 2018.

[24] D. Goldsmith and M. Rees, *The End of Astronauts: Why Robots Are the Future of Exploration*. Harvard University Press, 2022.

[25] O. Firschein, *Artificial Intelligence for Space Station Automation: Crew Safety, Productivity, Autonomy, Augmented Capability*. Noyes Publications, 1986.

[26] S. Rukh, *Artificial Intelligence in Space Exploration*. Independently Published, 2023.

[27] D. Damayanti, E. Royani, and J. Gultom, *The Law of Building Artificial Intelligence Robots in Space: Missions and Planetary Exploration with Solar Power Technology*. LAP LAMBERT Academic Publishing, 2023.

CHAPTER 3 ROBOTICS IN SPACE EXPLORATION

"Robotics is not about making machines to serve us. It's about creating machines that can be our partners."

—Cynthia Breazeal

3.1 INTRODUCTION

Robotics is the discipline of designing and constructing machines, called robots.

A robot is an autonomous mechanical device that is designed to sense its environment, carry out computations to make decisions, and perform actions like humans in the real world. It is a system that contains sensors, control systems, power supplies, and software, all working together to perform a task. Robotics is a relatively young field with highly ambitious goals. It is producing a huge range of devices, from autonomous vacuum cleaners to military drones [1]. Robotics has also allowed us to explore the unknown and expand our knowledge of the universe without risking human lives. It is opening up new possibilities for exploration and discovery in space. Space exploration has always been exciting for humans as it ignites our curiosity about the mysteries of the universe. Science fiction has become reality: we are going to the Moon and Mars. An example of a man on the moon is depicted in Figure 3.1 [2].

Figure 3.1 A man on the moon [2].

The role of robotics in space exploration is crucial because robots can perform tasks that are too dangerous or impractical for humans. Robots can also extend the reach and duration of human missions, by providing support, assistance, and companionship. They make it possible for humans to explore the unknown without risking their lives.

Space exploration has long been the domain of daring astronauts and sophisticated spacecraft. Robotics plays a crucial role in space exploration by enabling the exploration of distant planets and celestial bodies that are too hazardous or difficult for humans to reach. Robots offer unparalleled resilience, unmatchable precision, and evolving autonomy, positioning them as indispensable tools in the future of space exploration. They have the ability to extend human capabilities and enable exploration beyond what humans can physically achieve. They have proven to be the indefatigable explorers of the cosmos [3].

In this chapter, we will delve into the fascinating roles of robots in space exploration.

The chapter begins with explaining what robots are. It describes space robotics. It presents some applications of robotics in space exploration. It highlights the benefits and challenges of robotics in space exploration. It covers space robots around the world. It predicts the future of robotics in space exploration. The last section concludes with comments.

3.2 WHAT ARE ROBOTS?

The word "robot" was coined by Czechriter Karel Čapek in his play in 1920. Isaac Asimov coined the term "robotics" in 1942 and came up with three rules to guide the behavior of robots and later added the zeroth law [4]:

- Law 0: A robot may not injure humanity or through inaction, allow humanity to come to harm.
- Law 1: Robots must never harm human beings,
- Law 2: Robots must follow instructions from humans without violating rule 1,
- Law3: Robots must protect themselves without violating the other rules.

Robots are becoming increasingly prevalent in almost every industry, from healthcare to manufacturing. Figure 3.2 indicates that robotics is one of the branches of artificial intelligence.

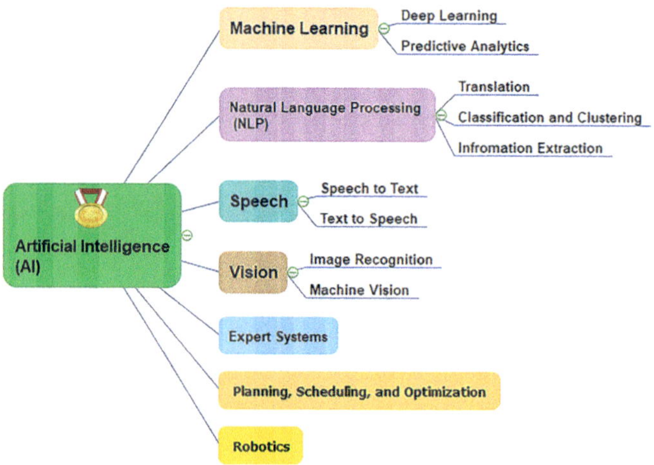

Figure 3.2 Robotics is one of the branches of artificial intelligence.

Although there are many types of robots designed for different environments and for different purposes/applications, they all share four basic similarities [5]:

(1) All robots have some form of mechanical construction designed to achieve a particular task;

(2) They have electrical components which power and control the machinery;

(3) All robots must be able to sense its surroundings; a robot may have light sensors (eyes), touch and pressure sensors (hands), chemical sensors (nose), hearing and sonar sensors (ears), etc.

(4) All robots contain some level of computer programming code.

Programs are the core essence of a robot since they provide intelligence. There are three different types of robotic programs: remote control, artificial intelligence, and hybrid. Some robots are programmed to faithfully carry out specific actions over and over again (repetitive actions) without variation and with a high degree of accuracy.

Robotics is an interdisciplinary field that involves the design, construction, operation, and use of robots. It is a branch of engineering and computer sciences that includes the design and use of machines that are capable of performing programmed tasks without human involvement. The field develops machines that can efficiently carry out various tasks, can automate tasks, and do various jobs that a human might not be able to do. Robots could someday be our drivers, companions, collaborators, teachers, specialists, and exploration pioneers.

3.3 SPACE ROBOTICS

The realm of space exploration has always captivated the human imagination, pushing the boundaries of what we know and understand about the universe. Robotics is a basic tool for space exploration, enabling us to study celestial bodies that would be inaccessible to humans and making significant contributions to our understanding of the universe. Space robotics has become an integral to the current state of space exploration.

Robotic systems are indispensable tools in space exploration, serving as our eyes and hands in the cosmos. Robotic space exploration involves the use of remotely operated or autonomous robots to explore celestial bodies and gather data in space. Space robots are specialized robots that replace humans in conducting scientific experiments, external vehicular activities, space exploration and other activities in space. These robots come in various forms. Among the diverse array of robots used in space exploration, rovers, satellites, and probes are the most prominent. These robotic systems play crucial roles in assembling, maintaining, and repairing infrastructure in space. Examples of robots used for space exploration include the following [6-8]:

- *Robotic Probes:* One of the most significant contributions of robotics to space exploration is the use of robotic probes. Probes are self-controlled devices sent on one-way trips to gather and transmit data about remote celestial bodies. Robotic probes and satellites have revolutionized our ability to explore celestial bodies beyond our own planet. These probes are unmanned spacecraft that are sent to explore the universe and gather data. For example, NASA's Voyager probes, launched in 1977, explored the outer solar system, and the data they collected about the planets and their moons has been invaluable in expanding our understanding of the universe. Robots could play a crucial role in exploring and preparing distant star systems for potential human colonization. By sending advanced robotic probes to these systems, we can gather information about their habitability and potential resources, paving the way for future human missions.

- *Robotic Rovers:* Another critical area where robotics has contributed to space exploration is the use of robotic rovers. These rovers are unmanned vehicles that are sent to explore the surface of other planets and moons. Rovers gather first-hand data about moons and planets by driving about the terrain and taking samples of a surface's crust. One of the most famous examples of a robotic rover is NASA's Mars rovers, which have explored the surface of Mars for over a decade. A robotic rover is shown in Figure 3.3 [8].

Figure 3.3 A robotic rover [8].

- *Space Robotic Dog:* Dogs will be returning to space, but this time in mechanical form. Scientists and students have deployed advanced robot dog explorers which might be on the moon in the future. With the ability to climb steep slopes, leap, dig, and explore elevated terrain, they could one day help humans explore the moon and other planets. Such a space robotic dog is shown in Figure 3.4 [9].

Figure 3.4 A space robotic dog [9].

- *Robotic Landers:* Robotic landers are robots designed to land on celestial bodies and remain stationary while conducting experiments and gathering samples. Similar to rovers, landers provide critical insights into the composition and structure of other worlds. Landers offer delicate descent and incredible stability, enabling them to touch down on uncharted territories, conducting in-depth examinations of soil, rock formations, and atmospheric conditions. Their findings have revolutionized our understanding of the geology of nearby planets and comets.

- *Robotic Orbiters:* They continuously circle celestial bodies like planets and moons, gathering critical data about objects beyond the reach of humans. For example, the observations of Mars Reconnaissance Orbiter and the Lunar Reconnaissance Orbiter contribute significantly to our understanding of Mars and the Moon. Their high-resolution cameras capture stunning images, revealing intricate surface features.
- *Robotic Manipulators:* These effectively act as robotic hands in space, extending human capabilities beyond our physical reach. They offer incredible reliability and dexterity, helping humans complete a wide range of tasks from a distance. . Their remarkable maneuverability and precision allow them to perform delicate operations, such as repairing spacecraft and satellites, collecting samples, and conducting intricate scientific investigations. The robotic arm on the Perseverance rover demonstrates the dexterity and precision of these robotic appendages. The Canadarm on the International Space Station is used to manipulate payloads and assist with spacewalks.
- *Robotic Assistants:* Robots are valuable in assisting astronauts during manned space missions. Robotic assistants, such as the Robonaut developed by NASA, are designed to carry out tasks that are too dangerous or complex for humans to perform. Robots can assist with maintenance, repairs, and even provide medical support to astronauts. By delegating certain tasks to robots, the workload on astronauts is reduced, allowing them to focus on critical mission objectives.
- *Robotic Space Telescopes:* Space telescopes, such as the Hubble Space Telescope, have transformed our understanding

of the universe. These telescopes are equipped with highly sensitive cameras and instruments that can capture images and data from distant galaxies and celestial objects.

- *Microgravity Robotics:* These refer to robots that are specifically designed to operate in zero-gravity environments, such as the International Space Station (ISS). These robots are typically used for maintenance and repair, scientific research, and to assist human astronauts with their daily tasks. One of the most vital roles of microgravity robotics is in the maintenance and repair of the ISS. Microgravity robotics also plays a crucial role in scientific research on the ISS. NASA, European Space Agency, and investors are pouring billions of dollars into creating robotics and AI applications suitable for zero-gravity conditions.
- *Commercial Space Robotics:* The growth of commercial space robotics adds an exciting dimension to the field, as private companies become increasingly involved in space exploration endeavors. The role of commercial space robotics is growing. Private companies are developing robotic spacecraft for various purposes, including satellite servicing, asteroid mining, and lunar exploration. This expansion of commercial activity is changing the landscape of space exploration.
- *Swarm Robotics:* Technologies like swarm robotics could allow multiple smaller robots to collaborate and complete complex tasks more efficiently than one large robot. Swarm robotics involves coordinating multiple robots to work together as a cohesive unit towards a common goal. This concept holds immense potential for space exploration as it allows for increased efficiency, redundancy, and adaptability. Imagine a swarm of small robots exploring the surface of a planet simultaneously, collecting data from various locations and sharing it with each other in real-time. If one robot were to encounter an issue or malfunction, the rest of the swarm could compensate for its loss, ensuring the mission's success.
- *Planetary Robot Systems:* These are robots that are designed

to operate on the surface of other planets, such as Mars or the Moon. These robots are typically much more complex than microgravity robotics and must be able to operate in harsh environments with extreme temperatures, radiation, and dust storms. Planetary robot systems also play a crucial role in preparing for human missions to other planets.

3.4 APPLICATIONS OF SPACE ROBOTICS

Robotics plays a crucial role in space exploration, enabling missions to gather data, conduct experiments, and explore environments beyond our planet. From autonomous exploration to repair and maintenance tasks, sample collection and analysis, and astronaut assistance, robotics is pushing the boundaries of what is possible beyond earth. Its broad applications include the following [11]:

- *Planetary Exploration:* Robots can orbit around planets, moons, asteroids, and comets, and collect data and images. Robots can also land on the surface of celestial bodies, and perform experiments, measurements, and sample collection. Some robots can drill, dig, or fly on the surface or in the atmosphere of other worlds. The practice of sending robots to space has not only optimized our exploration capabilities but also paved the way for future interplanetary missions

- *Space Exploration Missions:* Robotic missions act as pioneers, paving the way for future human exploration and colonization of space. Robots can enable and support human missions to space, by providing transportation, communication, navigation, and safety. Robots can be deployed ahead of human missions to prepare habitats, gather resources, and conduct preliminary experiments. They provide essential data and insights about celestial bodies, helping scientists and engineers understand the challenges and opportunities that await human explorers. They can assist the astronauts in maintaining and operating the space station, by performing routine or complex tasks, such as repairs, inspections, or experiments. Sample return missions represent another exciting aspect of space exploration. These missions involve not only the deployment of robotic systems to collect samples but also the complex task of returning these samples to Earth for in-depth analysis. The search for extraterrestrial life is one of the primary motivations behind space exploration. Robotic missions investigate environments where life might exist, examining the potential for water, organic molecules, and energy sources.

- *Space Science:* Scientific discoveries are at the core of space exploration. Robotic missions have made numerous scientific discoveries. Robots can advance the scientific knowledge and technological capabilities of space exploration, by conducting experiments, testing new systems, and demonstrating new concepts. Robots have played a crucial role in space exploration and research, enabling us to push the boundaries of our understanding of the universe.
- *Autonomous Exploration:* This is one of the most significant and exciting contributions of robotics in space exploration. With advancements in artificial intelligence (AI) and machine learning algorithms, space probes and rovers can navigate and explore celestial bodies with minimal human intervention. Robots will become increasingly capable of making decisions independently based on real-time data. This autonomy will enable them to explore distant planets or moons more efficiently without relying heavily on human operators. For example, NASA's Mars rovers Spirit and Opportunity have already demonstrated remarkable autonomy by conducting scientific experiments without constant human guidance.
- *Repair and Maintenance:* Spacecraft repair and maintenance are critical aspects of long-duration missions. In the harsh environment of space, repair and maintenance of spacecraft and satellites are incredibly challenging tasks. However, robotics has provided a solution to this problem. Robots equipped with specialized tools and robotic arms can perform repairs and maintenance tasks on spacecraft. This eliminates the need for human astronauts. Robotic manipulation systems in space have become indispensable assets, enabling a range of activities from assembly and maintenance to intricate

repairs and experiments.

- *In-Space Assembly:* In-space assembly missions play a critical role in advancing space exploration and ensuring the longevity of space infrastructure. Advanced robotics is revolutionizing in-space assembly and maintenance missions.
- *Data Collection:* Robotic technology has also revolutionized the process of sample collection and analysis in space exploration. Robotic probes can collect detailed scientific data on planetary surfaces, including soil composition, atmospheric conditions, and potential signs of life, providing valuable insights into the universe. For example, the Mars rovers, Curiosity and Perseverance, have been instrumental in gathering valuable data about the Red Planet. This data provides invaluable insights into the composition, atmosphere, and geological features of planets, moons, asteroids, and comets.

3.5 BENEFITS

The use of space robots has played a significant role in advancing scientific pursuits since the commencement of space exploration. Robots offer a practical solution to the myriad challenges posed by space exploration. We can send robots to explore space without having to worry so much about their safety. Despite being expensive to make, robots are cheaper than humans to deploy into space. The robots can do different tasks that aid astronauts and the science industry. Space robots also play a crucial role in testing the viability of technologies intended for human missions. Other benefits of space robots include [11]:

- *Cost-effectiveness:* Robotic missions in space also offer increased efficiency and cost-effectiveness compared to manned missions. Robots are cheaper to build, launch, and operate than human missions, as they do not require life support systems, food, water, or oxygen. The use of robots for space exploration is becoming increasingly attractive as they can work autonomously and do not require life support systems, making them more cost-effective than human astronauts.

- *Durability:* Robots can withstand harsh conditions like extreme temperatures, radiation, and vacuum, allowing them to explore planets like Mars, Venus, and Jupiter's moons where humans cannot survive. They can also operate for longer periods of time, and cover larger distances, than humans.
- *Flexibility:* Robots can be designed and customized for specific purposes, such as roving, flying, drilling, or sampling.
- *Creativity:* Robots can inspire and stimulate human curiosity, imagination, and innovation, by providing new perspectives, insights, and challenges. They can also foster collaboration and cooperation among different disciplines, sectors, and nations, by sharing data, resources, and goals.
- *Decision-making:* Advanced AI and sensor technology allow robots to navigate complex terrains, detect interesting features, and make independent decisions during exploration, optimizing mission efficiency.
- *In-situ Experiments:* Robots can perform scientific experiments on other planets by collecting samples, analyzing them on-site, and transmitting data back to earth.
- *Reducing Risk:* Space exploration is inherently risky, with astronauts facing numerous challenges and dangers. By sending robots instead of humans to explore distant celestial bodies, we can significantly reduce the risk to human life. Robotic systems have significantly enhanced the efficiency and safety of space exploration missions. Sending humans into space is not only costly but also poses numerous risks.
- *Human Support:* On the International Space Station, robotic arms and automated systems assist astronauts with tasks like maintenance, assembly, and experiments, freeing up their time for more complex research.
- *Collaboration:* International collaboration is a hallmark of space exploration. Space agencies and organizations around the world work together to advance our understanding of the universe. While robots have proven their worth in

space exploration, they are not meant to replace humans entirely. The synergy between humans and robots is vital for successful missions. Space exploration often involves international collaboration. For example, the Mars rovers are the result of collaborations between NASA and international partners. The collaboration between NASA and international partners has led to the success of various missions.

3.6 CHALLENGES

Space robotics face unique challenges, such as the need for autonomous navigation due to communication delays, extreme temperature variations, and exposure to cosmic radiation. The vast distances in space introduce communication delays that necessitate autonomous navigation by the robotic systems. Designing robots to traverse the surface of the Moon or Mars presents specific challenges. Other challenges of space robots include the following [9]:

- *Limitations:* Robots are still limited by their hardware, software, and communication capabilities, compared to humans. They may not be able to perform tasks that require intuition, judgment, emotion, or creativity.
- *Delays:* Robots rely on commands and signals from earth, which can take minutes or hours to reach them, depending on the distance. This can cause delays, lags, or gaps in the communication and control of the robots, and reduce the responsiveness and interactivity of the mission.
- *Ethics:* While robotics offers immense potential, ethical dilemmas arise. Robots raise ethical and social questions, such as who owns, controls, and benefits from them, and who is responsible, accountable, and liable for them. They may also pose threats or challenges to the environment, security, or sovereignty of space, and the rights and interests of other actors, such as humans, animals, or aliens.
- *Emotions:* Robots may not be able to evoke or elicit the same emotions, feelings, or sensations, as humans. They may also lack the human touch, connection, or empathy,

that can enhance the quality and meaning of the exploration experience.

- *Space Debris:* Space debris is becoming a major problem, with thousands of more satellites on track to launch through the end of this decade. Some companies and agencies are starting to test ways to clean it up. Space debris includes satellites that could be repaired or taken out of orbit. Many of these objects are tumbling, which makes rendezvous and docking with them a challenge. Figure 3.5 shows some space debris [12].

Figure 3.5 Some space debris [12].

3.7 SPACE ROBOTS AROUND THE WORLD

NASA and the European Space Agency (ESA) have deployed various types of robotic systems for space exploration. Nations across the globe have been running on the race to explore space for a century now. As nations send more rockets and satellites to the outer space, they get more information about the floating components. We consider how the following nations have incorporated robotics in their space exploration.

- *United States:* NASA's collaboration and funding of advanced robotics development will almost certainly benefit the economy at large in unpredictable ways. NASA's Perseverance rover, currently exploring Mars, is a prime example of how robots are aiding in the search for life beyond Earth. NASA is continuing its efforts to develop more robots that can be sent to outer space. It is also developing humanoid robots, such as shown in Figure 3.6 [9]. This robot, known as Valkyrie, later became part of NASA's Space Robotics Challenge.

Figure 3.6 A humanoid robot [9].

- *United Kingdom:* Several universities in UK support academic effort in robots. For example, University of Oxford is known globally for its academic excellence offering top-tier programs in AI and robotics. The university's robotics courses are focused on autonomous systems and human-robot interaction, preparing students for research and industry roles. Imperial College's commitment to research is reflected in its strong output of publications and its close collaborations with industry partners.
- *China:* China's Chang'e lunar exploration program, named after the Chinese moon goddess, has been a significant player in the field of lunar exploration. The program, initiated by the China National Space Administration (CNSA), has been a series of increasingly ambitious missions to understand the moon and its resources. The Chang'e 5 mission, launched in November 2020, marked another milestone for China's lunar exploration. It was designed to collect lunar soil and rock samples and return them to earth. In spite a few technical

challenges, the mission was a success, making China the third country to return samples from the moon.

- *India:* Chandrayaan-1, launched by the Indian Space Research Organization (ISRO) in October 2008, was India's first lunar probe and a significant milestone in the country's space exploration journey. It used a Polar Satellite Launch Vehicle (PSLV) to enter a high, polar orbit around the Moon. Its success demonstrated India's technological capabilities and laid the groundwork for future lunar explorations. SPACE India is committed to nurturing the next generation of space science students by providing unparalleled educational standards. Their passion for inspiring and mentoring students is evident in their unwavering dedication to cultivating young minds and fostering a love for space science.

3.8 FUTURE OF ROBOTICS IN SPACE EXPLORATION

As we stand on the precipice of a new era in space exploration, the role of robots has never been more crucial. The role of robotics in space exploration has been substantial, and it is expected to continue to grow in the coming years. As technology continues to evolve, the role of robotics in space exploration will become more significant. Future developments in robotics could lead to the creation of more advanced robots that can explore even more challenging environments, such as the subsurface oceans of Europa. With the development of more advanced robots, we can expect to learn even more about the universe and the mysteries it holds.

The future of robotics in space exploration holds immense potential. With the development of more advanced AI algorithms, robotic systems will become even more autonomous and capable of performing complex tasks. This opens up the possibility of sending robots to explore distant planets and moons, where the extreme conditions make human presence impractical or impossible. As the field continues to advance, we can expect even more exciting breakthroughs in the future. As we set our sights on more ambitious space exploration endeavors like manned missions to Mars or establishing a lunar base, robotics will play an even more crucial role. As we delve deeper into the 21st century, the realm of robotics continues to expand its horizons, transforming industries and pushing boundaries beyond our wildest imaginations. As technological advancements continue to propel the field of robotics, their role in space exploration will continue to expand. The future of space explorations is in the hands of AI and robotics. As shown in Figure 3.7, the space exploration programs are adopting Artificial Intelligence (AI) and robotics to fast-track their mission [11].

Figure 3.7 Space exploration programs are adopting AI and robotics [11].

3.9 CONCLUSION

Robotics plays a crucial role in space exploration by extending human capabilities, reducing risks, enhancing efficiency and safety in missions, and paving the way for future human exploration. It has already had a profound impact on space exploration and research, enabling us to explore distant planets, study celestial objects, and repair and maintain space-based assets. Robots are pioneering space exploration by acting as the primary explorers in challenging environments, utilizing their resilience, precision, and growing autonomy to venture into areas inaccessible to humans.

Space exploration is more exciting and complex than ever as we strive to reach the stars. Numerous opportunities have been made possible by the desire for Mars missions, lunar exploration, asteroid mining, and satellite servicing. Space robotics has had numerous remarkable contributions to our understanding of the universe. As technology progresses, we can expect more sophisticated robotic systems that can autonomously navigate and perform intricate operations with minimal human intervention. The role of space robotics in exploration and innovation is expected to expand tremendously. More information on robotics in space exploration can be found in the books in [13-26] and the following related journals:

- *Robotica*
- *Robotics*
- *Robitics and Autonomous*
- *Robotics and Computer-Integrated Manufacturing*
- *Advanced Robotics*
- *Autonomous Robots*
- *Journal of Robotics*
- *Journal of Robotic Systems*
- *Journal of Robotic Surgery*
- *Journal of Robotics and Mechatronics*
- *Journal of Intelligent & Robotic Systems*
- *Journal of Mechanisms and Robotics-Transactions of the ASME*
- *Journal of Automation, Mobile Robotics and Intelligent Systems*
- *Journal of Future Robot Life*
- *IEEE Robotics and Automation Letters*
- *IEEE Transactions on Robotics*
- *International Journal of Medical Robotics and Computer Assisted Surgery*
- *International Journal of Robotics Research*
- *International Journal of Social Robotics*

- *International Journal of Humanoid Robotics*
- *International Journal of Advanced Robotic Systems*
- *Science Robotics*
- *Soft Robotics*
- *Military Review*
- *Journal of Military Learning*
- *NCO Journal*
- *Space Policy*

REFERENCE

[1] M. N. O. Sadiku, U. C. Chukwu. A. Ajayi-Majebi, and S. M. Musa, "A primer on robotics, " *International Journal of Trend in Scientific Research and Development,* vol. 6, no. 7, November-December 2022, pp. 614-621.

[2] "Man on the moon – Neil Armstrong's iconic photograph," August 2017,

https://amateurphotographer.com/iconic-images/moon-iconic-photograph-neil-armstrong/

[3] M. N. O. Sadiku, U. C. Chukwu, and J. O. Sadiku, "Space robotics," *Horizon: Journal of Humanities and Artificial Intelligence*, vol. 2, no. 8, 2023, pp. 63-71.

[4] "Human–robot interaction," *Wikipedia*, the free encyclopedia

https://en.wikipedia.org/wiki/Human–robot_interaction

[5] "Robotics," *Wikipedia*, the free encyclopedia

https://en.wikipedia.org/wiki/Robotics

[6] S. Mahendra, "Robots in space," August 2023,

https://www.aiplusinfo.com/blog/robots-in-space/

[7] 'How robots are pioneering space exploration," November 2023,

https://www.captechu.edu/blog/how-robots-are-pioneering-space-exploration

[8] A. Khede, "Space Robotics Market Trends: Exploring the Future of Space Exploration," November 2023,

https://www.linkedin.com/pulse/space-robotics-market-trends-exploring-future-apurva-khede-cmx6c

[9] E. E. Matuszak, "5 Robots pushing the frontiers of space exploration," August 2023,

https://knowhow.distrelec.com/defence-aerospace-and-marine/5-robots-pushing-the-frontiers-of-space-exploration/

[10] "The role of robotics in space exploration,"

https://fastercapital.com/topics/the-role-of-robotics-in-space-exploration.html

[11] "The future of space explorations is in the hands of AI and robotics," January 2021,

https://www.analyticsinsight.net/artificial-intelligence/the-future-of-space-explorations-is-in-the-hands-of-ai-and-robotics

[12] B. Weir, "Nearly 30,000 objects are hurtling through near-Earth orbit. That's not just a problem for space," February 2024,

https://www.cnn.com/2024/02/21/climate/space-debris-solution-climate-scn/index.html

[13] M. N. O. Sadiku, *Robotics and Its Applications*. Moldova, Europe: Lambert Academic Publishing, 2023.

[14] R. L. Thomas, *Revolutionary Robots in Space*. Lerner Publications TM, 2019.

[15] Y. Wang, *Space Robotics*. Springer Nature, 2020.

[16] H. Toth, *Robotics in Space (Space Exploration)*. Brightpoint Press, 2022.

[17] R. Nagelhout, *Space Robots (Robots and Robotics)*. Powerkids Press, 2016.

[18] K. W. Larson, *Space Robots (Robotics in Our World)*. Amicus, 2017.

[19] F. Ehlers, *Design Space Exploration in Robotics (Studies in Systems, Decision and Control, 229)*. Springer, 2025.

[20] W. T. Huntress and M. Y. Marov, *Soviet Robots in the Solar System: Mission Technologies and Discoveries (Space Exploration)*. Springer, 2011.

[21] Y. Gao (ed.), *Space Robotics and Autonomous Systems: Technologies, Advances and Applications*. Institution of

Engineering and Technology, 2021.

[22] T. Kanade and Y. Xu, *Space Robotics: Dynamics and Control*. Springer, 2012.

[23] R. Wiens, *Red Rover: Inside the Story of Robotic Space Exploration, from Genesis to the Mars Rover Curiosity*. Basic Books, 2013.

[24] A. A. Desrochers (ed.), *Intelligent Robotic Systems for Space Exploration*. Springer, 2012.

[25] E. Zereik (ed.), *Space Robotics Supporting Exploration Missions: Vision, Force Control and Coordination Strategies*. United Kingdom: Nova Science Pub Inc., 2013

[26] D. Goldsmith and M. Rees, *The End of Astronauts: Why Robots Are the Future of Exploration*. Harvard University Press, 2022.

CHAPTER 4
DRONES IN SPACE EXPLORATION

"I think the greatest threat to the privacy of Americans is the drone, and the use of the drone and the very few regulations that are on it today, and the booming industry of commercial drones."

—Dianne Feinstein

4.1 INTRODUCTION

The vast expanse of outer space has long captured the imagination of humanity. Our solar system, including the sun and everything that orbits it (planets, asteroids, moons, comets, and meteoroids), has attracted more attention from various space agencies compared to other cosmological systems. Various space agencies have invested resources into exploring our solar system physically and through observation. There has been an interest in studying our neighboring planets and moons, such as Venus, Mars, and Titan. Progress in recent technologies has enabled space drones to be considered as valuable platforms for planetary exploration. Due to the advantages of drones compared to other approaches in planetary exploration, ample research has been carried out by different space agencies in the world, including NASA, to apply drones in other solar bodies.

Drones are increasingly being used in space exploration, transportation, and other applications. They are reshaping our understanding of the universe and revolutionizing the space industry by making it more cost-effective, efficient, and accessible. Drones provide an unprecedented capability to traverse difficult terrains and access remote locations, enabling the exploration of regions that were previously inaccessible to surface rovers or landers.

Drones are being utilized across the globe for various diverse uses by commercial, state, military, and individual users. In recent years, there has been a tendency to design and develop concepts of drones and robotic systems for planetary exploration. The integration of drones into space transportation and exploration represents a paradigm shift in how we approach the cosmos. Drones can make space more accessible, cost-effective, and efficient. Scientists use them to complement rovers and other observatory equipment to collect space data. Drones are expected to play a key role in the future of space exploration [1].

This chapter explores major applications of drones in space exploration. It begins with describing what a drone is. It covers space drones. It presents some applications of drones in space exploration. It highlights some benefits and challenges of drones in space exploration. The last section concludes with comments.

4.2 WHAT IS A DRONE?

The FAA defines drones, also known as unmanned aerial vehicles (UAVs), as any aircraft system without a flight crew onboard. Drones include flying, floating, and other devices, including unmanned aerial vehicles (UAVs), that can fly independently along set routes using an onboard computer or follow commands transmitted remotely by a pilot on the ground. A typical drone is shown in Figure 4.1 [2]. A drone is usually controlled remotely by a human pilot on the ground, as typically shown in Figure 4.2 [3]. Drones can range in size from large military drones to smaller drones. Drones, previously used for military purposes, have started to be used for civilian purposes since the 2000s. Since then, drones have continued to be used in intelligence, aerial surveillance, search and rescue, reconnaissance, and offensive missions as part of the military Internet of things (IoT). Today, drones are used for different purposes such as aerial photography, surveillance, agriculture, entertainment, healthcare, transportation, law enforcement, etc.

Figure 4.1 A typical drone [2].

Figure 4.2 A drone is usually controlled by operators on the ground [3].

Commercial drones have come a long way in the last decade. Drones work much like other modes of air transportation, such as helicopters and airplanes. When the engine is turned on, it starts up, and the propellers rotate to enable flight. The motors spin the propellers and the propellers push against the air molecules downward, which pulls the drone upwards. Once the drone is flying, it is able to move forward, back, left, and right by spinning each of the propellers at a different speed. Then, the pilot uses the remote control to direct its flight from the ground [4].

Drone laws exist to ensure a high level of safety in the skies, especially near sensitive areas like airports. They also aim to address privacy concerns that arise when camera drones fly in residential areas. These include the requirement to keep your drone within sight at all times when airborne. In the United States, drones weighing less than 250g are exempt from registration with civil aviation authorities. If your drone exceeds 250g in weight, you will also require a Flyer ID, which requires passing a test [6]. It is necessary to register as an operator, be trained as a pilot, and have civil liability insurance, in addition to complying with various flight regulations, and those of the places where their use is permitted.

Most drones have a limited payload, usually under 11 pounds. Drones are classified according to their size. Here are the different drone types:

- Nano Drone: 80-100 mm
- Micro Drone: 100-150 mm
- Small Drone: 150-250 mm
- Medium Drone: 250-400 mm
- Large Drone: 400+ mm

One of the emerging trends in drone use for factories is the utilization of LiDAR technology. LiDAR stands for Light Detection and Ranging. This technology provides accurate depth information essential for understanding the three-dimensional structure of the environment. LiDAR sensors emit laser beams to measure distances to objects, creating high-resolution 3D maps of the surrounding terrain and objects. The ability to capture detailed data through LiDAR technology has opened up opportunities for better predictive maintenance, reduction in inspection times, and overall cost savings [5].

4.3 SPACE DRONES

Scientists need special tools and devices to explore space. Drones have emerged as useful tools, needed in space to observe and survey areas. Drones help scientists learn more about asteroids as well as other planets and moons. The advantage of using drones is they are cheaper than sending people and have better potential to fly over places the rovers cannot access. A standard drone cannot fly to space independently for many reasons. For the propellers to effectively lift the drone, there must be enough air in the atmosphere. As you move up the Earth's atmosphere, the air becomes thinner and the pressure drops. The reasons drones cannot fly in space include the following [7]:

(1) Lack of air push.

(2) Lack of movement.

(3) Lack of powerful engine.

(4) Drones cannot work in vacuum.

(5) Propellers cannot work.

That is why NASA is working on a different propulsion system for the drones meant to fly to space. Thus, drones can fly in space but with customization to withstand low atmospheric pressure and weak gravity. A modified or assisted drone can fly to space, although the pilots have to add modified communication equipment to effectively control the drone. To fly in space, the drone needs some assistance in the form of a weather balloon, which allows the drone to get to the incredibly high altitude required in order to be able to see the circumference of planet earth. Once the weather balloon reaches a certain point, the balloon will pop, resulting in the drone falling back to earth.

Space drone is a spacecraft designed to dock with existing satellites that are running low on fuel but are otherwise operational. It uses electric propulsion to maneuver to its target satellite, attaching to the satellite's launch vehicle interface ring via robotic arms.

Solar panels and energy storage systems are helping space drones sustain longer missions. Drones are designed and fabricated with different shapes and sizes and hence various methods and materials for manufacturing. A typical space drone is shown in Figure 4.3 [8].

Figure 4.3 A typical space drone [8].

4.4 APPLICATIONS

In recent years, there has been a paradigm shift in the way we approach space transportation and exploration, with the integration of drones playing a pivotal role. Drone technology has multiple capabilities and applications. Space drones can assist in mapping, sample collection, and geological and atmospheric analysis. Common applications of drones in space exploration include the following [9]:

- *NASA:* NASA is using drone technology for a variety of reasons, such as aerial imagery, inspections, and mapping. Drones are used at NASA's Stennis Space Center to capture images and video and record data. NASA is developing a nuclear rotorcraft drone called Dragonfly that will explore the skies and surface of Titan. At NASA's Stennis Space Center at Mississippi, drones are becoming a go-to resource for use on difficult and potentially dangerous jobs, helping to save time and costs. As NASA integrates drone mapping operations into existing software, future scanning is expected to allow the creation of 2D or 3D real-time maps and models. A team of NASA engineers wants to put drones on Mars.

NASA already uses drones in many of their operations. Figure 4.4 shows an astronaut using a digital glove in space to control a drone on the moon [9].

Figure 4.4 A astronaut uses a digital glove in space to control a drone on the moon [10].

- *Aerospace:* Drones are transforming the aerospace industry in many ways. Drones can inspect aircraft, runways, and airport terminals more quickly, precisely, and cost-effectively than traditional methods. For example, Airbus uses drones to photograph the upper part of its A330 aircraft during final inspections. Drones can send service work orders to maintenance teams as soon as they identify a fault. This can reduce maintenance costs and improve safety. The aerospace industry is also exploring the use of space drones to provide inspections of commercial aircraft. With traditional technology, visual inspections can take up to six hours, but space drones could significantly reduce the amount of time and provide increased accuracy and ease of documentation [11]. Figure 4.5 shows a drone flying at the Golden Triangle Regional Airport [12].

Figure 4.5 A drone flying at an airport [12].

- *Satellite Communications:* Satellite communication can provide reliable wireless connectivity for drones almost anywhere in the world. Drones with satellite communication can provide real-time data and imagery to command centers, allowing for better situational awareness and decision-making. They can be used for efficient delivery, inspection, and monitoring services in areas such as e-commerce, logistics, and agriculture. They can also be used for collaborative exploration and experimentation in science, engineering, education, and art. However, satellite communication terminals can be bulky, making them unsuitable for smaller drones. Figure 4.6 shows drones crossing with satellite [13].

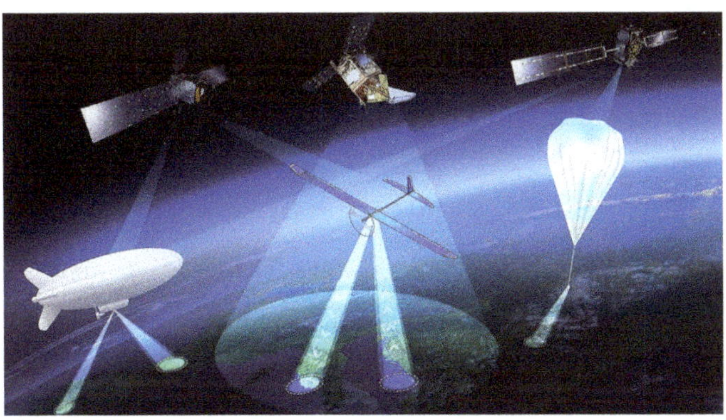

Figure 4.6 Drones crossing with satellites [13].

- *Safety Inspection:* The safety of astronauts and the integrity of space vehicles are paramount. Drones can be used for safety inspections before and during launch. They can check for debris on the launchpad, ensuring a clean and safe departure. Drones equipped with high-resolution cameras and sensors can be used to inspect launch facilities. Additionally, drones can monitor the weather conditions, providing real-time data that helps in making critical launch decisions.
- *Autonomous Spacecraft Repairs:* Space exploration missions are often marred by the challenges of long distances and communication delays. Drones can be deployed to autonomously repair spacecraft, extending their operational lifetimes. In the future, autonomous drones could repair critical systems, such as solar panels or communication antennas, while the spacecraft is far from Earth. Autonomous drones could be equipped with tools and technology to handle routine maintenance and repairs.
- *Aerial Exploration:* On planets like Mars, drones are being used to conduct aerial exploration. For example, the Mars Helicopter has provided stunning imagery and data from the Martian surface, showcasing the potential for drones to reach areas that are inaccessible to traditional rovers.
- *Sample Collection:* Drones can also be used to collect samples from planets and moons. A drone can fly to a specific location, gather samples, and return to the main spacecraft, all while operating autonomously.
- *Resource Survey:* Drones can be deployed to conduct resource surveys on the Moon. They can explore the lunar surface, identify valuable resources like water ice, and even perform preliminary mining activities. This approach could significantly reduce the cost of lunar missions by utilizing local resources.

4.5 BENEFITS

Drones have the potential benefits in terms of cost-efficiency, versatility, and adaptability to different environments. Safety is a significant benefit of drone usage. Drones offer unique capabilities to get close views of potentially life-threatening situations. They can impact almost any business and provide a positive return-on-investment quickly. They are rapidly evolving beyond capturing data into major transportation vehicles and carrying us into their future. Drones can also be autonomous, and some can withstand harsh conditions like extreme temperatures or high levels of radiation. They can play a role in deploying satellites and in maintaining space stations. Other benefits of drones in space exploration include the following [14]:

- *Sample Collection:* Drones can be used to collect samples from planets and moons. A drone can fly to a specific location, gather samples, and return to the main spacecraft, all while operating autonomously. This could be a game-changer for missions aiming to study the composition of extraterrestrial bodies.

- *Planetary Exploration:* Exploring other planets and celestial bodies has always been a fascination for scientists and space agencies. Drones are playing a pivotal role in these endeavors, bringing new dimensions to planetary exploration. They can explore planets like Mars and Venus, and can provide high-resolution information about the planet's surface, atmosphere, and interior.

- *Asteroid Exploration:* Drones can explore and analyze asteroids up close to provide data that helps us understand the early solar system. They can also help identify suitable mining targets.

- *Aerial Exploration:* On planets like Mars, drones are being used to conduct aerial exploration. The Mars Helicopter, also known as Ingenuity, is a prime example. It has provided stunning imagery and data from the Martian surface, showcasing the potential for drones to reach areas that are inaccessible to traditional rovers.

- *Survey:* Drones can be deployed to conduct resource surveys on the Moon. They can explore the lunar surface, identify valuable resources like water ice, and even perform preliminary mining activities. This approach could significantly reduce the cost of lunar missions by utilizing local resources.
- *Infrastructure Development:* Drones can assist in the development of lunar infrastructure. They can be used to transport materials and equipment, construct habitats, and even set up power generation systems. This paves the way for sustainable lunar colonization.
- *Infrastructure Inspection:* Spacecraft and rockets are stored and prepared in massive hangars and launchpads. Drones equipped with high-resolution cameras and sensors can be used to inspect these facilities, ensuring that they are in optimal condition. They can identify issues such as structural damage, leaks, and electrical faults that could jeopardize a mission.
- *Safety Inspections:* The safety of astronauts and the integrity of space vehicles are paramount. Drones can perform safety checks before and during a launch to ensure a safe departure. They can check for debris on the launchpad and monitor weather conditions. They can check for debris on the launchpad, ensuring a clean and safe departure. Additionally, drones can monitor the weather conditions, providing real-time data that helps in making critical launch decisions.

4.6 CHALLENGES

There are challenges and limitations associated with drone technology in space exploration. These include issues related to power supply, materials, communication delays, thermal control, and the need for robust navigation systems in unfamiliar terrains. Organizations like NASA have been trying their best to send drones at the heights, but there are many factors which pulled them down. Drones which are used on Earth cannot be used the same way in space, because we cannot push air over there. It is challenging for an astronaut to control a drone while outside of their ship on the moon's surface. Other challenges include the following [15]:

- *Regulations:* Space is subject to various international treaties and agreements. Space drones must adhere to regulatory frameworks, which can be complex to navigate. Regulations in the industry have helped address safety and security concerns. Drones have been operating under what amount to temporary regulations, created by governing bodies that did not yet have enough data to create fully fleshed-out rules. For example, the FAA treats all UAVs the same under its regulations, from small drones marketed at hobbyists to large, complex UAVs that can transport large parcels over great distances. The main aim of these regulations is to prioritize the safety of the airspace being traveled by traditional manned commercial and private aircraft.
- *Harsh Environments:* Space is a hostile environment with extreme temperatures, radiation, and vacuum conditions. Drones must be designed to withstand these conditions, which can be damaging to both hardware and electronics.
- *Latency:* In planetary exploration, drones often operate on celestial bodies with significant communication delays, making real-time remote piloting challenging. They must operate autonomously to some degree.
- *Complex Navigation:* The navigation and mapping of alien landscapes are complex tasks. Drones need advanced sensors, cameras, and processing capabilities to avoid obstacles and explore efficiently.
- *Limited Payload Capacity:* Space drones must be lightweight, which often limits their payload capacity for scientific instruments or tools.
- *Power Constraints:* Power generation and energy storage are crucial for space drones. Solar panels, nuclear power sources, or other innovative solutions are required for extended missions.
- *Reliability:* The reliability of space drones is paramount, given the costs and risks associated with space missions.

Redundancy and fail-safe mechanisms are necessary.

- *Autonomy:* Space missions, especially planetary exploration, can last for months or even years. Drones must be autonomous and resilient over long durations.

Further research and development in this field is needed to overcome these challenges. Figure 4.7 shows the pros and cons of drone technology [16].

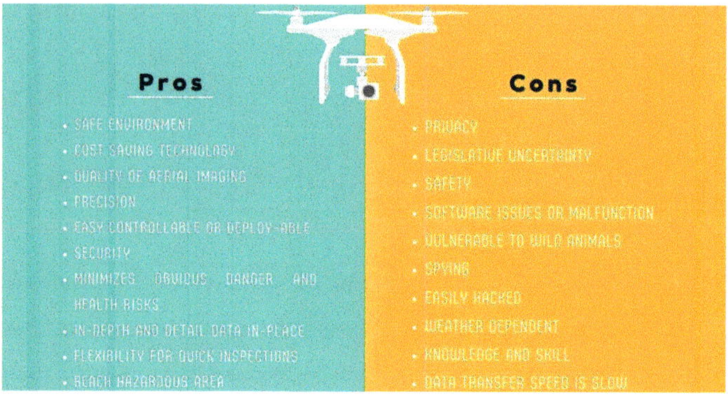

Figure 4.7 The pros and cons of drone technology [16].

4.7 CONCLUSION

As technology continues to advance, we can only expect the role of drones in space exploration to expand further. Drones have the potential to revolutionize future space missions, enabling greater understanding and exploration of the universe. They have proven to be indispensable tools in space exploration and transportation. With their ability to access hard-to-reach areas, conduct autonomous repairs, and gather valuable data, drones are poised to continue transforming our dreams of exploring the cosmos into reality. Although only one drone has been sent to Mars, more research is underway to create better crafts for space exploration.

Drones are a transformative technology that will change in ways we likely never imagined. Much of the future of drones may seem like sci-fi but it i much closer than you may think. Using drones to explore the other planets or moons is one of the main priorities of space agencies. NASA already uses drones in many of their operations, and they hope to use drones widely in the future. More information about drones in space can be found in the books in [17-20] and in the following related journals:

- *Drones*
- *Vertical Space e-Magazine*
- *Progress in Aerospace Sciences*

REFERENCE

[1] M. N. O. Sadiku, U. C. Chukwu, and J. O. Sadiku, "Drones in space," *Innovative Mult-disciplinary Journal of Applied Technology*, vol. 2, no. 10, 2024, pp. 43-49.

[2] T. Bishop, "FAA looking into picture taken by drone above Space Needle," December 2014,

https://www.geekwire.com/2014/faa-looking-picture-taken-drone-space-needle/

[3] "The 5 best surveillance drones: Next-level inspection UAVs,"

https://www.zdnet.com/article/best-surveillance-drone/

[4] "How drones work and how to fly them," May 2024,

https://dronelaunchacademy.com/resources/how-do-drones-work/

[5] "What are the main applications of drones?" June 2024,

https://www.jouav.com/blog/applications-of-drones.html

[6] "Drones in manufacturing: A game-changer for industry,"

https://viper-drones.com/industries/infrastructure-drone-use/manufacturing/#:~:text=The%20integration%20of%20drones%20into,on%20manufacturing%20is%20no%20exception.

[7] "Drones in space? Is it possible? Let's find out," March 2020,

https://medium.com/srmscro/drones-in-space-is-it-possible-lets-find-out-1031fd726f87

[8] https://www.vectorstock.com/royalty-free-vector/drone-outer-space-earth-vector-8604522

[9] "The use of drones in space transportation and exploration,"

https://www.av8prep.com/aviation-library/part-107-drone/the-use-of-drones-in-space-transportation-and-exploration#:~:text=Drones%20can%20also%20be%20 used,the%20composition%20of%20extraterrestrial%20bodies.

[10] “Astronaut to use a digital glove in space to control a drone on the moon,’ November 2019,

https://dronevideos.com/astronaut-to-use-a-digital-glove-in-space-to-control-a-drone-on-the-moon/#:~:text=As%20the%20hand%20and%20fingers,or%20closed%20the%20hand%20is.

[11] “Space drones and the future of drones in aerospace,” June 2018,

https://www.proponent.com/news/future-of-drones-in-aerospace/#:~:text=The%20aerospace%20industry%20is%20also,accuracy%20and%20ease%20of%20documentation.

[12] B. Guillot and M. Dowell, “Airport benefits of drone technology,” April 2019,’

https://www.aviationpros.com/aircraft/unmanned/article/12436848/airport-benefits-of-drone-technology

[13] “Crossing drones with satellites: ESA eyes high-altitude aerial platforms,” November 2017,

https://www.esa.int/Applications/Satellite_navigation/Crossing_drones_with_satellites_ESA_eyes_high-altitude_aerial_platforms#:~:text=Crossing%20drones%20with%20satellites%3A%20ESA%20eyes%20high%2Daltitude%20aerial%20platforms,-28%2F11%2F2017&text=ESA%20is%20co-nsidering%20extending%20its,link’%20between%20drones%20and%20satellites.

[14] “The use of drones in space transportation and exploration,”

https://www.av8prep.com/aviation-library/part-107-drone/the-use-of-drones-in-space-transportation-and-exploration#:~:text=Sample%20Collection,the%20composition%20of%20extraterrestrial%20bodies.

[15] “Challenges and innovations in space-related drone applications,”

https://www.av8prep.com/aviation-library/part-107-

drone/challenges-and-innovations-in-space-related-drone-applications#:~:text=Challenges%20in%20Space%2DRelated%20Drone%20Applications&text=Harsh%20Environments%3A%20Space%20is%20a,to%20both%20hardware%20and%20electronics.

[16] https://equinoxsdrones.com/10-major-pros-cons-of-unmanned-aerial-vehicleuav-drones/

[17] D. R. Faust, *Drones in Space (Drones Are Everywhere!)*. PowerKids Press, 2019.

[18] R. M. Marx, *Creating Space: Drones, Just War, and Jus Ad Vim*. Kent State University, 2016.

[19] M. London, Space Drones. Abdo Publishing, 2021.

[20] *Drones in Space: Proceedings of the 32nd Annual Wisconsin Space Conference*, 2023

CHAPTER 5
BIG DATA IN SPACE EXPLORATION

"Hiding within those mounds of data is knowledge that could change the life of a patient, or change the world."

—Atul Butte

5.1 INTRODUCTION

The world has seen a digital revolution where more and more work is being conducted online. Storing this activity has led to the concept of big data, i.e., large datasets that are otherwise difficult to manage. Because of these characteristics, big data requires new technologies and techniques to capture, store, and analyze. The cloud word for big data is shown in Figure 5.1 [1]. Typical sources of big data are shown in Figure 5.2 [2].

Figure 5.1 The cloud word for big data [1].

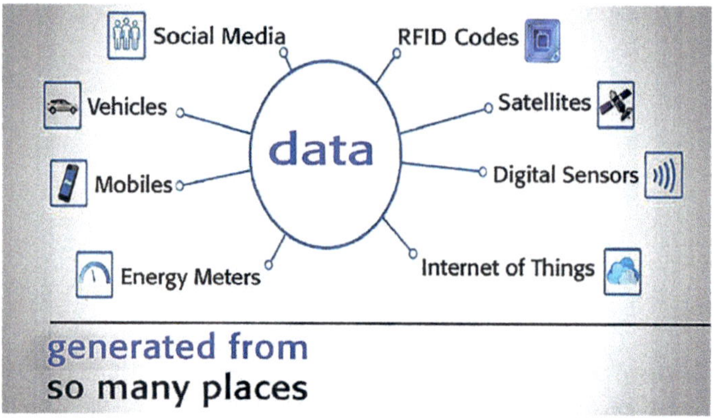

Figure 5.2 Typical sources of big data [2].

Big data plays a central role in space exploration, where satellites, telescopes, rovers, and space probes collect and analyze colossal volumes of data from space and celestial bodies. It empowers the transformation of raw data into transformative knowledge. It meticulously sifts through terabytes of information, uncovering patterns, predicting celestial events, and optimizing missions. It is about pushing the boundaries of human understanding and propelling us further into the cosmic abyss [3].

Space exploration has always been a primary goal for humanity, pushing the boundaries of our understanding and opening up new frontiers. Space exploration, once guided primarily by telescopes and human curiosity, now relies heavily on data analytics to decipher the secrets of the cosmos. Figure 5.3 depicts the journey through the history of space data [4].

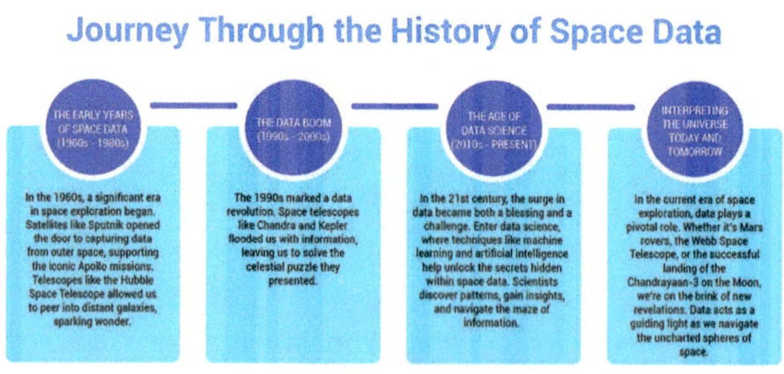

Figure 5.3 Journey through the history of space data [4].

Data analytics is important in contemporary space missions,

encompassing data collection, processing, analysis, and interpretation. It plays a fundamental role in spacecraft design and trajectory optimization by employing mathematical models, simulations, and optimization techniques. It supports communication, navigation, and lunar exploration through techniques like error correction codes, encryption, and machine learning. It is essential for tasks such as filtering, noise reduction, and anomaly detection, ensuring data integrity, and instrument reliability during processing.

Big data plays a crucial role in space exploration by enabling scientists to analyze vast amounts of data collected from telescopes, satellites, and space probes. The role of big data in space exploration is undeniable, revolutionizing the way we observe, analyze, and understand the universe. Big data analysis influences space exploration by enabling us to better understand space data, and unlock the mysteries of the universe. Big data analytics has become an indispensable tool for space agencies and researchers worldwide. Utilizing big data analytics techniques, astronomers are now able to process immense datasets and uncover patterns and trends that would otherwise remain hidden [5].

In this chapter, we will delve into the fascinating role of big data in space exploration. The chapter begins with explaining what big data is all about. It presents some characteristics of big data. It describes big data space exploration. It covers some applications of big data in space exploration. It highlights the benefits and challenges of big data in space exploration. The last section concludes with comments.

5.2 WHAT IS BIG DATA?

Big data applies to data sets of extreme size (e.g. exabytes, zettabytes) which are beyond the capability of the commonly used software tools. It involves situation where very large data sets are big in volume, velocity, veracity, and variability [6]. The data is too big, too fast, or does not fit the regular database architecture. It may require different strategies and tools for profiling, measurement, assessment, and processing.

Big Data is essentially classified into three types [7]:

- *Structured Data:* This is highly organized and is the easiest to work with. Any data that can be stored, accessed, and processed in the form of fixed format is known as a structured data. It may be stored in tabular format. Due to their nature, it is easy for programs to sort through and collect data. Structured data has quantitative data such as age, contact, address, billing, expenses, credit card numbers, etc. Data that is stored in a relational database management system is an example of structured data.
- *Unstructured Data:* This refers to unorganized data such as video files, log files, audio files, and image files. Any data with unknown form or the structure is classified as unstructured data. Almost everything generated by a computer is unstructured data. It takes a lot of time and effort required to make unstructured data readable. Examples of unstructured data include Metadata, Twitter tweets, and other social media posts.
- *Semi-structured Data:* This falls somewhere between structured data and unstructured data, i.e., both forms of data are present. Semi-structured data can be inherited such as location, time, email address, or device ID stamp.

The different types of big data are depicted in Figure 5.4 [8].

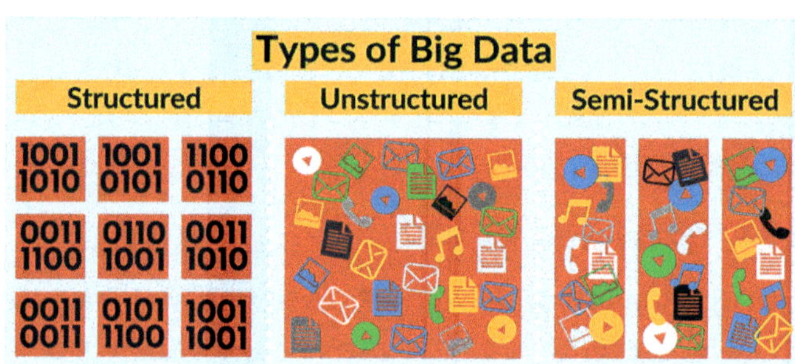

Figure 5.4 Types of big data [8].

The process of examining big data is often referred to big data analytics. It is an emerging field since massive computing capabilities have been made available by e-infrastructures [9]. Analytics include statistical models and other methods that are aimed at creating empirical predictions. Data-driven organizations use analytics to guide decisions at all levels. Several techniques have been proposed for analyzing big data. These include the HACE theorem, cloud computing, Hadoop, and MapReduce [10].

5.3 CHARACTERISTICS OF BIG DATA

Big data is growing rapidly and expanding in all science and engineering, including physical, biological, and medical services. Different companies use different means to maintain their big data. As shown in Figure 5.5 [11], big data is characterized by 42 Vs. The first five Vs are volume, velocity, variety, veracity, and value [2].

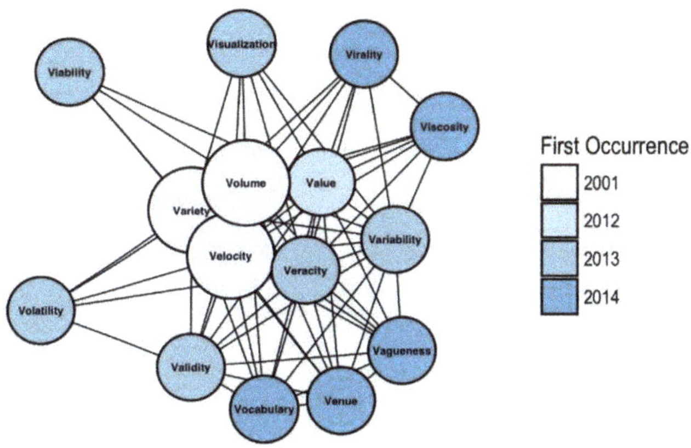

Figure 5.5 The 42 V's of big data [11].

- *Volume:* This refers to the size of the data being generated both inside and outside organizations and is increasing annually. Some regard big data as data over one petabyte in volume.
- *Velocity:* This depicts the unprecedented speed at which data are generated by Internet users, mobile users, social media, etc. Data are generated and processed in a fast way to extract useful, relevant information. Big data could be analyzed in real time, and it has movement and velocity.
- *Variety:* This refers to the data types since big data may originate from heterogeneous sources and is in different formats (e.g., videos, images, audio, text, logs). BD comprises

of structured, semi-structured or unstructured data.

- *Veracity:* By this, we mean the truthfulness of data, i.e. weather the data comes from a reputable, trustworthy, authentic, and accountable source. It suggests the inconsistency in the quality of different sources of big data. The data may not be 100% correct.
- *Value:* This is the most important aspect of the big data. It is the desired outcome of big data processing. It refers to the process of discovering hidden values from large datasets. It denotes the value derived from the analysis of the existing data. If one cannot extract some business value from the data, there is no use managing and storing it.

On this basis, small data can be regarded as having low volume, low velocity, low variety, low veracity, and low value. Additional five Vs has been added [12]:

- *Validity:* This refers to the accuracy and correctness of data. It also indicates how up to date it is.
- *Viability:* This identifies the relevancy of data for each use case. Relevancy of data is required to maintain the desired and accurate outcome through analytical and predictive measures.
- *Volatility:* Since data are generated and change at a rapid rate, volatility determines how quickly data change.
- *Vulnerability:* The vulnerability of data is essential because privacy and security are of utmost importance for personal data.
- *Visualization:* Data needs to be presented unambiguously and attractively to the user. Proper visualization of large and complex clinical reports helps in finding valuable insights.

Instead of the 10V's above, some suggest the following 5V's: Venue, Variability, Vocabulary, Vagueness, and Validity) [13].

Industries that benefit from big data include the healthcare, financial, airline, travel, restaurants, automobile, sports, agriculture, and hospitality industries. Big data technologies are playing an essential role in farming: machines are equipped with sensors that measure data in their environment. Structured and unstructured data are generated in various types [14-16].

5.4 BIG DATA SPACE EXPLORATION

Space exploration has always been a realm of discovery, pushing the boundaries of human knowledge and technology. It has always been at the forefront of human ambition, pushing the boundaries of our understanding and opening up new frontiers. The vast unknown of space has captivated humanity for millennia, but unlocking its secrets demands more than celestial gazing.

The sheer volume of data has presented a new challenge for astronomers. How to process and analyze this massive amount of information effectively is not an easy task. Processing and making sense of big data from space is a monumental task. This is where big data comes into play. Big data analysis plays a crucial role in the study of dark matter and dark energy, which together constitute around 95% of the universe. Analyzing big data from space has transformed the field of astronomy, enabling scientists to unlock the secrets of the universe.

Today, data analytics in space industry plays a pivotal role, acting as the bridge between raw information and profound knowledge. Data is the backbone of space exploration, and with advancements in technology, we are now generating an unprecedented amount of data. From analyzing planetary formation data to predicting space weather patterns, big data plays a crucial role in virtually every aspect of space exploration. Data analytics has become a pivotal tool, transforming how we explore, understand, and utilize the cosmos. It enables continuous monitoring of astronaut health, analyzing biometric data to detect potential health issues, optimize life support systems, and ensure astronaut safety. It enhances deep space communication, optimizing data transmission and reducing communication delays. Data analytics in space exploration is illustrated in Figure 5.6 [17].

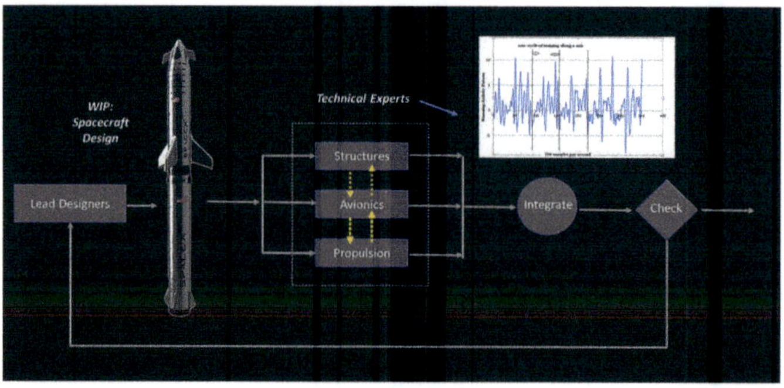

Figure 5.6 Data analytics in space exploration [17].

More than 50 years ago, space exploration was a race to the moon between the US and the former Soviet Union. In 1998, the International Space Station launch marked a highlight for international cooperation in space between the U.S., Russia, Japan, Canada and Europe. Today, more countries like as India and China are pursuing their own goals in space, which could challenge the US as a world leader. Americans continue to see an essential role for the United States as a leader in space exploration. Most Americans agree that it is essential for US to be a leader in space exploration.

NASA is engaged in a wide range of activities in space, including exploration and applied and basic research. Since its creation in 1958, NASA has achieved some pretty impressive feats of science. When NASA was created, it had to invent the technology to get where we needed to go, and we will continue to push the boundaries of technology into the future. New emerging technologies that open opportunities for research and exploration with minimal investments include NASA's satellites, such as typically shown in Figure 5.7 [18].

Figure 5.7 Satellites for exploration [18].

There is far less public urgency for NASA to send humans to the moon or Mars and to search for other planets that could support life. Americans by and large view NASA as critical to space exploration [19]. NASA collects, stores, processes, and analyzes data. It uses a system called the Mission Data Processing and Control System (MPCS) to manage and process data. The application for the NASA's Mars Curiosity rover uses production data and creates an easy way for engineers to monitor instruments and measurements faster and more efficiently. Elasticsearch permits handling large amounts of data from Curiosity's sensors, such as the temperature on the Martian surface. Figure 5.8 shows the artist's concept of NASA's X-57 'Maxwell' aircraft [20], while Figure 5.9 displays NASA's gateway in lunar orbit [20].

Figure 5.8 The artist's concept of NASA's X-57 'Maxwell' aircraft [20].

Figure 5.9 NASA's gateway in lunar orbit [20].

The X-57 will be the first all-electric X-plane and will be flown to demonstrate the benefits that electric propulsion may yield for the future of aviation. NASA will continue to be a global leader in scientific discovery. It has continued to push the boundaries of knowledge to deliver on the promise of American ingenuity and leadership in space. Its future will continue to be a story of human exploration, technology, and science.

5.5 APPLICATIONS OF BIG DATA SPACE EXPLORATION

Big data analytics in space exploration unlocks patterns within volumes of observations to advance astrophysics, guide future missions, and reveal our universe's deepest mysteries. The cosmos may be vast, but through data, it becomes knowable. Scientists and engineers use big data for everything from predicting weather on earth to monitoring ice caps on Mars to searching for distant galaxies. Here are some ways that big data is used in space exploration [21]:

- *Astronomy:* Astronomy has always been a data-intensive science. Data science has significantly transformed the field of space exploration, particularly in the realm of astronomy. By leveraging data science techniques, researchers are able to tackle complex astrophysical problems, uncovering hidden patterns and shedding light on the universe's origin, evolution, and large-scale structure. Data science techniques can help researchers understand the universe's origin, evolution, and structure. For example, machine learning algorithms can classify galaxies and their mergers, while image recognition algorithms can identify celestial objects.
- *Predicting Space Weather:* Space weather can impact satellites, spacecraft, and even power grids on earth. Big data analysis plays a crucial role in predicting space weather patterns by analyzing vast amounts of data from satellites, ground-based observatories, and space weather models. The predictions enable space agencies and operators to take proactive measures to protect spacecraft and astronauts and mitigate the effects of space weather on critical infrastructure. Space exploration models predict space weather events that could severely impact satellites and astronauts in space. Crucial data-driven analytics will forecast perilous space weather to ensure mission success.
- *Astronauts' Health:* Radiation exposure poses significant risks to astronauts' health during space missions, especially during long-duration flights beyond Earth's protective

magnetosphere. Big data analysis of radiation monitoring data collected from spacecraft, space stations, and dosimeters worn by astronauts helps scientists assess exposure levels and mitigate health risks. This data-driven approach enables space agencies to optimize mission planning, crew scheduling, and spacecraft design to ensure the safety and well-being of astronauts.

- *Earth Observation:* Data analytics plays a crucial role in interpreting satellite data for earth observation. The Sun's activity, characterized by phenomena such as solar flares, coronal mass ejections, and sunspot cycles, influences space weather conditions in the solar system. Satellites play a crucial role in monitoring and studying earth's environment, climate, and natural resources from space. Big data analysis of satellite imagery and remote sensing data provides valuable insights into changes and trends in earth's surface, atmosphere, and oceans. By analyzing multi-spectral imagery, radar data, and other earth observation datasets, scientists can track deforestation, urbanization, pollution, and other environmental changes over time. Information derived from this analysis supports decision-making in areas such as land use planning, disaster response, and natural resource management. Big data analysis of cosmic ray data collected by ground-based detectors and space-based observatories helps scientists study the origins, propagation, and interactions of cosmic rays. Free earth-observation datasets from the European Space Agency, NASA and the US Geological Survey have helped organizations around the world develop object-detection platforms.

- *Space Telescope Data:* Modern telescopes equipped with sophisticated sensors and cameras capture detailed images of celestial objects. They generate an astronomical amount of data. Space telescopes equipped with advanced instrumentation enable astronomers to search for signs of extraterrestrial life beyond our solar system. For example, the Hubble Space Telescope has collected over 150 terabytes of data throughout its mission. Analyzing this vast amount of data manually would be an impossible task. Big data analysis of space telescope data, including spectroscopic observations and transit surveys, helps scientists identify potentially habitable exoplanets and study their atmospheres for biosignatures.

5.6 BENEFITS

Many analysis tools offer fast analysis and visualization of data, and have proven to be very valuable. They help space probes make faster and better decisions, reduce failure, and improve life on planet earth. Big data analysis of data collected from space probes, telescopes, and simulations allows researchers to identify patterns and trends that reveal the underlying mechanisms of planetary formation. Data analytics enhances deep space communication, optimizing data transmission, and reducing communication delays. As space missions grow more complex, data scientists play a vital role in ensuring success. One of the significant advantages of data science in astronomical research is the ability to unveil the secrets of the universe. Other benefits include [21]:

- *Decision Making:* From mission design to resource allocation, data science in space industry, empowers informed decision-making throughout exploration endeavors, maximizing efficiency, minimizing risks, and paving the way for future discoveries. Space agencies are using big data tools to rapidly analyze their data and make time-efficient decisions. Data science helps make informed decisions throughout space exploration, from mission design to resource allocation. The ability to tell stories with data is paramount for inspiring decision makers.
- *Farming:* The Climate Corporation uses their satellite data to enable farmers all around the globe to find more sustainable ways to grow substantially more food. Farmers can use image data to better understand what factors affect the growth of crops. There are factors that can be detected from space, such as weather patterns, exposure to sunlight, air quality or pest activity.
- *Monitoring:* Spacecraft monitor everything from our

home planet to faraway galaxies, beaming back images and information to earth. Spacecraft operates in extreme conditions, making health monitoring a critical aspect of space missions. Wearable sensors and advanced analytics are transforming how astronaut health is monitored during space missions. By continually tracking vital signs, biometrics, and even microbial flora in the body, subtle shifts can trigger earlier medical alerts than relying on scheduled tests or self-reported issues alone.

- *Autonomous Systems:* Data science is instrumental in developing autonomous systems for space exploration. Autonomous robots and spacecraft rely on data analysis to make informed decisions and adapt to dynamic environments. By integrating data science techniques with robotics, scientists can develop intelligent systems capable of exploring distant planets.
- *Simulation:* Data science helps in simulating and modeling complex astronomical scenarios, allowing astronomers to test theories and hypotheses. These simulations aid in understanding celestial processes and predicting the behavior of astronomical objects under different conditions.
- *Communication:* Data analytics improves communication reliability and data quality by analyzing signal patterns, noise, and interference. Releasing public domain images of a lonely blue planet or footage of astronauts on spacewalks elegantly communicate their accomplishments to the public while inspiring future engineers and scientists.
- *Mission Planning:* Predicting celestial events becomes more accurate through the use of data analytics. Data analytics helps optimize mission trajectories and improve spacecraft design by analyzing past mission data, current space weather conditions, and predictive models.
- *Data Processing:* Data analytics transforms raw data from on board instruments into meaningful information using signal and image processing, data mining, and pattern recognition. Data analytics in space exploration is pivotal for processing

the enormous volumes of data transmitted back to earth by space probes, satellites, and telescopes. Big data analytics techniques are essential for managing, processing, and extracting meaningful information from this data deluge.

- *Data Quality:* Astronomy heavily relies on observational data to study celestial objects. By applying data analytics techniques, filtering out noise and extracting relevant information enhances the quality of data. Data science improves the precision and accuracy of observational data collection.
- *Improved Efficiency:* By automating data analysis processes, astronomers can focus more on the interpretation of results and the formulation of new theories. This leads to a significant increase in efficiency and productivity.
- *Space Photography:* Photographs are universal means of communication. Over time, photography techniques grew from pointing a camera out of a window to using powerful mirrors and telescopes to capture impossibly distant subjects. To share these images with the world, NASA needed to solve photographic and data challenges. The space photography is crucial to communicating the mission of NASA. From the beginning, NASA understood that space photography played a vital role in its missions. The data from satellites and cameras is captured and stored locally.
- *Space Tourism:* Most adults in US expect that people will routinely travel in space as tourists in the next 50 years. More than half expect space tourism to become routine. However, Americans are not enthusiastic about traveling to space themselves.

5.7 CHALLENGES

Data captured via space satellites pose significant challenges and risks, such as data quality and reliability, data security and privacy, data ownership and access, data ethics and fairness, and data regulation and compliance. Data transmission in space is still a challenge and finding a way to ease that process would be beneficial for many space missions. Some challenges associated with data analytics in space exploration include data accuracy, privacy, ethical use of AI, and making AI algorithms compatible with harsh space environments. Space missions generate enormous volumes of data, presenting significant challenges in data storage, processing, and analysis. Other challenges include [22,23]:

- *Space Debris:* As Americans look to the future of space, a large share expect problems with human-made debris. There will be a major problem with debris in space from rockets, satellites, and other human-made objects. Ratings are more mixed when it comes to the job private companies are doing to limit debris from objects like rockets and satellites. Figure 5.10 shows some space debris [24].

Figure 5.10 Space debris [24].

- *Safety:* Space debris poses a growing threat to spacecraft and satellites in orbit. Big data analysis of space debris tracking data helps space agencies and operators predict debris trajectories and assess collision risks.

- *Communication:* Communicating from the ground station to a spacecraft is a complex challenge, largely due to the extreme distances involved. When data are transmitted and received across thousands or millions of miles, the delay and potential for disruption or data loss is significant. Transferring data also becomes more challenging because the signal gets weaker due to distances involved. Advanced communication technologies are required to enhance deep space exploration for both robotic and human missions. Optical communications technologies can dramatically improve communications between spacecraft and earth far better than radio communications.

- *International Collaboration:* Space exploration and scientific research are inherently collaborative endeavors that require cooperation among nations and space agencies worldwide. Big data analysis of space data often involves international collaboration, with researchers sharing data, resources, and expertise to address complex scientific questions and challenges. International collaborations enable scientists to tackle fundamental questions in astronomy, planetary science, and space physics more effectively.
- *Risk:* Whenever an astronaut steps out of a space vehicle, it is an incredible, high-level-risk activity. We need to critically think about the risk to human life and the value return for assigning humans to do activities. Smart autonomous systems have the capability to improve the efficiency and speed while minimizing the risk and cost involved with planetary missions.
- *Data Quality:* How can we ensure that the data collected in space are accurate, dependable, and consistent? How can we deal with the noise, interference, and distortion that may affect the data transmission? How can we validate, verify, and correct the data?
- *Data Security:* How can we protect the data from unauthorized access, modification, or deletion? How can we prevent cyberattacks, hacking, or sabotage that may compromise the data or the spacecraft? How can we encrypt, authenticate, and backup the data?
- *Data Availability:* How can we ensure that the data are accessible and usable by the intended users? How can we overcome the limitations of bandwidth, latency, and storage that may affect the data transfer? How can we manage the

data lifecycle, from collection to disposal?

- *Data Usability:* How can we make the data understandable and meaningful for the users? How can we standardize, harmonize, and integrate the data from various sources, formats, and systems? How can we analyze, visualize, and share the data?

5.8 CONCLUSION

Space data is about to get big. Big data has revolutionized the field of space technology, enabling us to explore the universe with unprecedented precision and efficiency. The integration of big data analytics has revolutionized the way we explore and understand the cosmos. It will play an increasingly vital role in shaping the future of space exploration. As technology evolves, the future holds even greater opportunities for data analytics in revolutionizing our understanding of the cosmos. By combining powerful data analytics techniques with increasingly vast amounts of data, astronomers are revolutionizing the way they conduct research, analyze celestial objects, and make groundbreaking discoveries.

Data analytics is not just transforming space exploration; it is redefining our understanding of the universe. This versatile tool has been a cornerstone of space missions, ensuring efficiency and success in the exploration of space and celestial bodies.

It provides astronomers with the tools and techniques necessary to analyze the massive volumes of data generated by telescopes and satellites. Data-driven intelligence provides the analytical foundation needed to realize humanity's boldest visions. As we continue to push the frontiers of space exploration, there is no doubt that big data will remain a cornerstone of astronomical research. More information about big data in space can be found in the books in [25-27].

REFERENCE

[1] R. Delgado, "The challenges of bringing BYOD to the military,"

https://socpub.com/articles/the-challenges-of-bringing-byod-to-the-military-11272

[2] J. Moorthy et al., "Big data: Prospects and challenges," *The Journal for Decision Makers*, vol. 40, no. 1, 2015, pp. 74–96.

[3] "10 Uses of data science in space exploration," March 2024,

https://medium.com/@analyticsemergingindia/10-uses-of-data-science-in-space-exploration-d958045213c2

[4] "Navigating uncharted space: Harnessing data analytics in space exploration," October 2023,

https://www.dasca.org/world-of-data-science/article/navigating-uncharted-space-harnessing-data-analytics-in-space-exploration

[5] M. N. O. Sadiku, U. C. Chukwu, and J. O. Sadiku, "Big data in space exploration," *Innovative Mult-disciplinary Journal of Applied Technology*, vol. 2, no. 10, 2024, pp. 227-239.

[6] M. N. O. Sadiku, M. Tembely, and S.M. Musa, "Big data: An introduction for engineers," *Journal of Scientific and Engineering Research*, vol. 3, no. 2, 2016, pp. 106-108.

[7] "The complete overview of big data,"

https://intellipaat.com/blog/tutorial/hadoop-tutorial/big-data-overview/

[8] R. Allen, "Types of big data | Understanding & Interacting with key types (2024),"

https://investguiding-com.custommapposter.com/article/types-of-big-data-understanding-amp-interacting-with-key-types

[9] P. Baumann et al., "Big data analytics for earth sciences: the earthserver approach," *International Journal of Digital Earth*,

vol. 19, no. 1, 2016, pp.3-29.

[10] X. Wu et al., "Knowledge engineering with big data," *IEEE Intelligent Systems*, September/October 2015, pp.46-55.

[11] "The 42 V's of big data and data science,"

https://www.kdnuggets.com/2017/04/42-vs-big-data-data-science.html

[12] P. K. D. Pramanik, S. Pal, and M. Mukhopadhyay, "Healthcare big data: A comprehensive overview," in N. Bouchemal (ed.), *Intelligent Systems for Healthcare Management and Delivery. IGI Global*, chapter 4, 2019, pp. 72-100.

[13] J. Moorthy et al., "Big data: Prospects and challenges," *The Journal for Decision Makers*, vol. 40, no. 1, 2015, pp. 74–96.

https://www.grandviewresearch.com/industry-analysis/industrial-wireless-sensor-networks-iwsn-market

[14] A. K. Tiwari, H. Chaudhary, and S. Yadav, "A review on big data and its security,"

Proceedings of IEEE Sponsored 2nd International Conference on Innovations in Information Embedded and Communication Systems, 2015.

[15] M. B. Hoy, "Big data: An introduction for librarians," *Medical Reference Services Quarterly*, vol. 33, no 3. 2014, pp. 320-326.

[16] M. Viceconti, P. Hunter, and R. Hose, "Big data, big knowledge: Big data for personalized healthcare," *IEEE Journal of Medical and Health Informatics*, vol. 19, no. 4, July 2015, pp. 1209-1215.

[17] T. Ebrahimi, " Dispatches from deep space: Imaging data and space photography," July 2022,

https://blog.westerndigital.com/space-photography/

[18] D. Kekare, "Data analytics in space exploration: Beyond earth's boundaries," March 2024,

https://www.linkedin.com/pulse/data-analytics-space-exploration-beyond-earths-durgesh-kekare-v5iwf#:~:text=Data%20analytics%20enhances%20deep%20space,mission%20control%20and%20scientific%20research.

[19] "Americans' views of space: U.S. role, NASA priorities and impact of private companies," July 2023,

https://www.pewresearch.org/science/2023/07/20/americans-views-of-space-u-s-role-nasa-priorities-and-impact-of-private-companies/

[20] "60 Years and counting - The future,"

https://www.nasa.gov/specials/60counting/future.html

[21] "Exploring the role of big data in space exploration," March 20 https://1ansah.in/blog/exploring-the-role-of-big-data-in-space-exploration/#:~:text=The%20role%20of%20big%20data%20in%20space%20exploration%20is%20undeniable,space%20agencies%20and%20researchers%20worldwide.24,

[22] D. Werner, "Big data, advanced algorithms and new approaches for space missions," October 2024,

https://spacenews.com/big-data-advanced-algorithms-and-new-approaches-for-space-missions/

[23] "Data governance in space — Key challenges and opportunities,"

https://www.cdomagazine.tech/opinion-analysis/data-governance-in-space-key-challenges-and-opportunities

[24] "Space debris: A threat just beyond the horizon," July 2022,

https://www.rfa.space/space-debris-a-threat-just-beyond-the-horizon/

[25] M. N. O. Sadiku, U. C. Chukwu, and P. O. Adebo, *Big Data and Its Applications*. Moldova, Europe: Lambert Academic Publishing, 2024.

[26] A. Agarwal, *Interplanetary Data Governance: Navigating the New Digital Frontier in Space*. Kindle Edition, 2024.

[27] F. Adam and P. Skoda (eds.), *Knowledge Discovery in Big Data from Astronomy and Earth Observation: Astrogeoinformatics*. Elsevier Science, 2020.

CHAPTER 6
CLOUD COMPUTING IN SPACE EXPLORATION

"Cloud is about how you do computing, not where you do computing."

—Paul Maritz

6.1 INTRODUCTION

Cloud computing is a means of pooling and sharing hardware and software resources on a massive scale. Users and businesses can access applications from anywhere in the world at any time. Companies offering these computing services are called cloud providers and typically charge for cloud computing services based on usage [1]. Some features of cloud computing are displayed in Figure 6.1 [2].

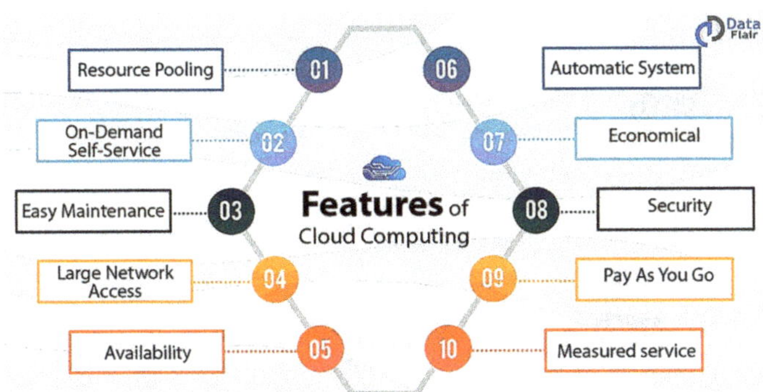

Figure 6.1 Some features of cloud computing [2].

The goal of cloud engineering is to create a highly scalable, flexible, and secure computing environment that can support the needs of modern businesses. Here on planet Earth, cloud computing has revolutionized the way we access and analyze data.

Space activity and exploration have a huge impact on daily life on earth. The space economy has seen significant advancements in recent years, with more companies exploring and commercializing space-related technologies. Space exploration is going through a major transformation period [3]. Cloud computing has emerged as a crucial factor in achieving success and progress for space exploration. It is used in space exploration for a variety of reasons, including data storage and management, data processing, access to information, communication, monitoring. Cloud service providers should target government organizations, satellite operators, and space exploration players that are keen to use cloud computing in their low-Earth orbit and lunar activities. The transition from the ISS to commercial space stations will generate new opportunities for partnerships between space industry players and cloud service providers [4].

Cloud computing is the on-demand delivery of computing resources, such as storage, processing power, and applications, via the Internet. It uses the Internet to quickly access, study, and read data over a shared server. Cloud technology is now transcending earth into space. In the space exploration, cloud computing plays a pivotal role in managing the vast amounts of data generated by satellites, telescopes, and other space-based assets. The adoption of cloud services by space industry players has increased significantly. These players have been making strategic moves to integrate space technology into their cloud services. Satellite-based cloud computing has the potential to revolutionize space exploration [5].

This chapter reviews the role cloud computing plays in the space exploration. It begins with discussing the basics of cloud computing. It covers cloud computing in space exploration. It presents some applications of cloud computing in space exploration. It highlights some benefits and challenges of cloud computing in space exploration. The last section concludes with comments.

6.2 CLOUD COMPUTING BASICS

Cloud computing represents a newly emerging service-oriented computing technology. It is the provision of scalable computing resources as a service over the Internet. It allows manufacturers to use many forms of new production systems such as 3D printing, high-performance computing (HPC), industrial Internet of things (IIoT), and industrial robots. It is transforming virtually every facet of modern manufacturing. It is innovating, reducing cost, and bolstering the competitiveness of American manufacturing [6].

The key characteristic of cloud computing is the virtualization of computing resources and services. Cloud computing is implemented in one of three major formats: software as a service (SAAS), platform as a service (PAAS), or infrastructure as a service (IAAS). These services are illustrated in Figure 6.2 [7] and explained as follows:

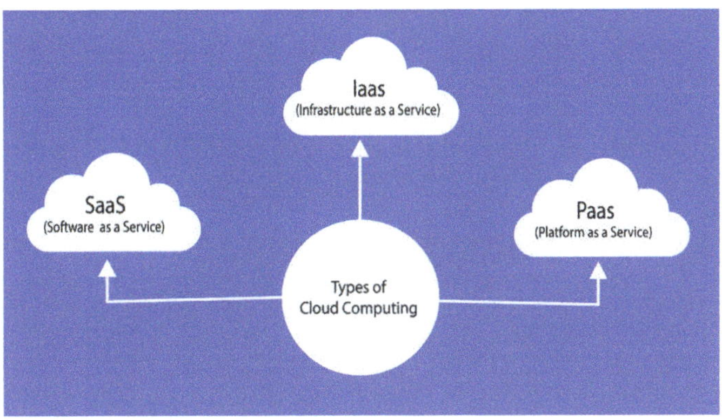

Figure 6.2 Three types of cloud computing services [7].

SaaS: This is a software delivery model in which software and associated data are hosted on the cloud. In this model, cloud service providers offer on-demand access to computing resources such as virtual machines and cloud storage. Nowadays oil & gas companies transition to cloud computing and implement SaaS solutions for operations.

PaaS allows the end-user to create a software solution using tools or libraries from the platform service provider. In this model, cloud service providers deliver computing platforms such as programming and execution.

In the IaaS model, cloud service providers can rent manufacturing equipment such as 3D printers.

Just like cloud computing, CM services can be categorized

into three major deployment models (public, private, and hybrid clouds) [8]:

- Private cloud refers to a centralized management effort in which manufacturing services are shared within one company or its subsidiaries. A private cloud is often used exclusively by one organization, possibly with multiple business units.
- Public cloud realizes the key concept of sharing services with the general public. Public clouds are commonly implemented through data centers operated by providers such as Amazon, Google, IBM, and Microsoft.
- Hybrid cloud that spans multiple configurations and is a composed of two or more clouds (private, community or public), offering the benefits of multiple deployment modes.
- These models are shown in Figure 6.3 [9]. Cloud computing finds application in almost every field

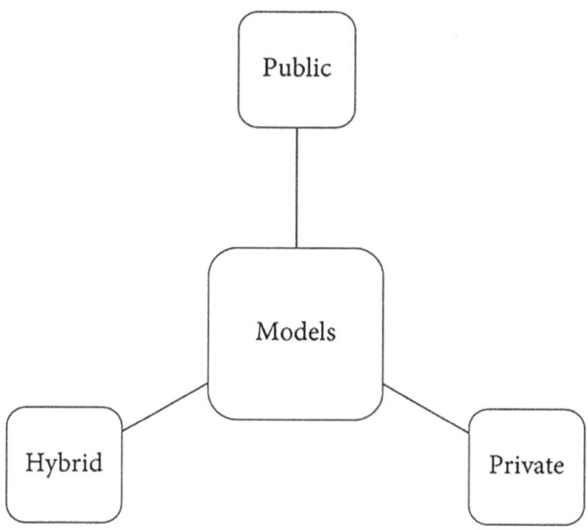

Figure 6.3 Cloud computing models [9].

6.3 CLOUD COMPUTING IN SPACE EXPLORATION

Space exploration is one of the most challenging and complex tasks humanity has ever undertaken. It requires advanced space technologies and techniques to overcome the obstacles it presents. One such technology that has the potential to revolutionize space exploration is satellite-based cloud computing. Satellite-based cloud computing can unlock the potential of space exploration by improving data processing and analysis, enabling autonomous operations, and enhancing collaboration and coordination. The management, security, and rapid exchange of an ever-increasing amount of data, much of it tactical, are crucial elements for the country's defense.

Cloud computing is making its way into space as cloud suppliers continue to push the frontiers of the technology. Satellite-based cloud computing refers to the use of satellites to provide cloud computing services, such as storage, processing, and software applications, to users on the ground. This can be useful in remote or underserved areas where traditional Internet infrastructure is lacking. The satellite-based cloud computing is still in the development phase, but it can potentially bring cloud computing capabilities to places where it was previously unavailable. One of the key benefits of satellite-based cloud computing is its ability to offload data to ground-based data centers for processing and analysis. Satellite-based cloud computing can also support autonomous operations on spacecraft [10].

6.4 APPLICATION OF CLOUD COMPUTING IN SPACE

The space economy has experienced significant growth in recent years, driven by advancements in technology, increased investment, and a growing interest in space exploration and commercialization. The possible applications of cloud tech to space travel are many. The following are typical applications:

- *Space Cloud:* This refers to utilizing satellites in low-earth orbit to provide cloud computing services. It is widely considered to be the next frontier in cloud computing, offering potential for significantly reduced latency, global reach, and enhanced data processing capabilities, particularly for applications requiring real-time data analysis and remote locations with limited terrestrial connectivity. Space cloud offers a unique solution to data management, along with unparalleled processing speed, flexibility, storage capacity and security on a global scale. The space cloud initiative integrates cloud infrastructure, AI, and supercomputing within a secure network of satellites orbiting the earth. Space cloud is expected to have a transformative influence on how governments and defense forces utilize and share data. Space cloud pushes the boundaries further to overcome limitations of speed, capacity, latency and security [11].

- *In-space Data Center:* This may be the next step for space cloud computing. Right now, everything is sent back to earth for storage and processing. In the future, data centers may be deployed in space to improve data transmission speed and processing power capacity. The commercialization of the low-Earth orbit (LEO) market and long-term lunar and space exploration missions will increase the use of data centers in space. The transition from the International Space Station (ISS) to commercial space stations are the main factors that will boost the use of in-space data centers. Figure 6.4 displays the Destiny Module inside the ISS, where astronauts perform many of their experiments [3].

Figure 6.4 The Destiny Module inside the ISS [3].

Applications related to earth observation, communications, science, and human spaceflight will drive the use of data centers in space. Government players drive most of the demand for in-space data centers in the science sector because they conduct most of the major scientific missions. R&D is required to overcome the major challenges regarding the implementation of in-space data centers.

- *Space Economy:* Cloud computing has become an integral part of the growing space economy. Numerous space economy businesses are currently using cloud computing for a diverse range of applications. By leveraging cloud-based platforms, space economy businesses can scale their operations, reduce costs, and drive innovation in space-related applications and services. In order to see the space cloud succeed, technologists must get innovative on how to efficiently transmit and store data collected from space probes and spacecraft. A space cloud would provide more efficient storage and transmission methods, so that power consumption would be reduced [12].
- *Satellite-based Cloud Computing:* Satellite-based cloud computing refers to the use of satellites to provide cloud computing services, such as storage, processing, and software applications, to users on the ground. This can be useful in remote or underserved areas where traditional Internet infrastructure is lacking. Data is sent to and from the satellites using radio frequency (RF) communications, and the cloud computing services are typically provided by data centers located on the ground. This satellite-based cloud computing is still in the development phase, but it can potentially bring cloud computing capabilities to places where it was previously unavailable. Thus, with the increasing number of satellites being launched into orbit, advances in technology, expansion of Internet connectivity in remote and underserved areas, and the growing demand for cloud computing services, the global satellite-based cloud computing technology is experiencing growth. Figure 6.5 shows space exploration using satellite [10].

Figure 6.5 Space exploration using satellite [10].

6.5 BENEFITS

Cloud computing enables real-time communication between the rover and mission control on earth, facilitating remote operation and data transmission. Less reliance on ground networks minimizes external influence, ensuring uninterrupted operations crucial for national security. Cloud-based services are finding their way into space to analyze satellite imagery and monitor space debris. Other benefits include [10,11]:

- *Decision Making:* Space cloud facilitates in-orbit decision-making, which includes reactive planning, scheduling decisions, and complex resource management within space networks. The Spaceborne Computer-2 on the International Space Station (ISS) is already using artificial intelligence, machine learning and cloud processing to make predictions and draw conclusions. Figure 6.6 shows the Spaceborne Computer-2 on the ISS [3].

Figure 6.6 The Spaceborne Computer-2 on the ISS [3].

- *Reduced Latency:* Processing data in orbit minimizes latency, facilitating real-time decision-making by reducing space-to-ground data transmissions.
- *Storage:* Storing data in space offers a secure backup solution that is less vulnerable to natural disasters, enhancing communication and data network resilience. Increased activity and the growing number of satellite constellations are driving the demand for data storage in space.
- *Accessibility:* Space-based cloud extends the reach of data processing capabilities to remote areas lacking the accessibility to traditional ground-based cloud infrastructure.
- *Security:* The space environment mitigates risks of cyberattack and physical tampering risks, ensuring secure data transmission and storage.
- *Situational Awareness:* Integrating data gathered and processed from satellites and unmanned aerial vehicles enables more comprehensive asset tracking and threat monitoring, enhancing situational awareness.
- *Autonomy:* Autonomous operations, such as taking images or adjusting the spacecraft's trajectory, require significant computing power and storage. These operations can be performed on a cloud-based infrastructure which allows for

real-time adjustments and decision making. This can greatly enhance the capabilities of spacecraft, allowing them to perform complex tasks with minimal human intervention.

- *Collaboration:* Cloud-based platforms foster collaboration between teams and organizations, streamlining the development of new space technologies. Satellite-based cloud computing plays a critical role in collaboration and coordination among space agencies and organizations. A centralized platform for collaboration can be provided through cloud computing, allowing team members to access and share data and resources easily. Engineers, scientists, and other professionals can work together seamlessly, accessing and sharing data from any location.
- *Cost-Effectiveness:* Satellite-based cloud computing can help to reduce the costs of space missions. By offloading data processing and analysis to ground-based data centers, spacecraft can more efficiently use their limited resources while also gaining access to more powerful computing resources.
- *Scalability:* Cloud computing enables organizations to scale their infrastructure according to their needs, providing a flexible environment for data storage and processing. As companies grow and their data requirements change, they can easily adjust their cloud resources, ensuring they can efficiently manage their operations.
- *Cost Optimization:* Cloud computing allows businesses to optimize costs by paying only for the resources they use. The pay-as-you-go model offered by cloud providers helps organizations better manage their budgets and avoid investing in expensive physical infrastructure.
- *Data Analysis:* Space economy companies generate vast amounts of data from various sources, such as satellites, telescopes, and space probes. Cloud computing facilitates the rapid analysis of this data by providing access to powerful data processing and analytics tools.

- *Space Debris:* Another area where cloud technology could be applied in space is the identification and management of space debris, which astronauts at the International Space Station have had to navigate around to prevent catastrophic collisions. Figure 6.7 shows an artist's conception of just some of the trillions of pieces of debris orbiting the Earth [13].

Figure 6.7 An artist's conception of pieces of debris orbiting the Earth [13].

6.6 CHALLENGES

There are data security challenges and huge power requirements, along with costs associated with data egress and other cloud services, which may be significant setbacks for the growth of cloud computing market. One thing that all satellites share is a long development and production cycle. These are complex machines that cannot easily be replaced. They take years to design, build, test and finally launch. Other challenges include [14]:

- *Data Collection:* There is the issue of data collection. Satellites produce huge quantities of data. At present, it takes days for their findings to be analyzed

- *Lack of Expertise:* Space industry players' lack of expertise regarding cloud computing/big data requirements along with the disconnect between the cloud computing and space industries limit the adoption of this technology.

- *Delay:* The constrained connectivity can delay communicating critical decisions to the astronauts, who often have to wait for information to reach ground control and be analyzed there and then returned with the necessary insight. That's hard enough with the space station, which orbits as high as 250 miles above the earth. Further mission into outer space will need stronger computing power at the astronauts' fingertips and a better pipeline for sharing information.

- *Astronauts' Healthcare:* Space travel may be profound, but it has a harsh impact on the human body. A major challenge addresses the needs of healthcare for astronauts on longer space missions. The effects on a human body of lengthy sojourns in space are not fully known, making technology that allows frequent monitoring of changes over time especially important. The farther astronauts travel away from earth, the more self-reliant they will need to be. That means their computer systems will have to process huge amounts of data locally, at the edge, to quickly answer questions and solve problems.
- *Next-Generation Computing:* Next-generation data computing for space travel is still under development, so there is more room for innovations. Even with improvements on the horizon, the quality of life and work for astronauts has greatly improved since the early days of exploration.

6.7 CONCLUSION

We are living in a new era in which data and software can be accessed anywhere with an Internet connection. Cloud computing is crucial for space exploration because all the new missions emerging in space in the next 10 or 20 years are going to require cloud-based technology. Governments and cloud service providers should work with space industry players to increase the use of in-space cloud computing. Engineers and researchers across companies should work together to improve communication and enabling experiments that will propel astronauts further into space while improving the lives of the earthbound as well. More information on the integration of cloud technology into space exploration is available from the books in [15-17].

REFERENCE

[1] M. N. O. Sadiku, S. M. Musa, and O. D. Momoh, "Cloud computing: Opportunities and challenges," *IEEE Potentials*, January-February 2014, pp. 34-36.

[2] "Features of cloud computing – 10 major characteristics of cloud computing,"

https://data-flair.training/blogs/features-of-cloud-computing/

[3] S. Ray, "Cloud 'data bursts' from space move astronauts closer to Mars — And improve life on earth," August 2021,

https://news.microsoft.com/source/features/innovation/cloud-data-bursts-from-space-move-astronauts-closer-to-mars-and-improve-life-on-earth/

[4] D. Christodoulopoulou, "LEO and lunar market growth to boost in-space cloud computing demand," September 2024,

https://www.kratosdefense.com/constellations/articles/leo-and-lunar-market-growth-to-boost-in-space-cloud-computing-demand

[5] M. N. O. Sadiku, P. A. Adekunte, and J. O. Sadiku, "Cloud computing in the space exploration," *International Journal of Trend in Scientific Research and Development*, vol. 8, no. 5, September-October 2024, pp. 1161-1167.

[6] S. Ezell and B. Swanson, "How cloud computing enables modern manufacturing," June 2017

https://itif.org/publications/2017/06/22/how-cloud-computing-enables-modern-manufacturing

[7] "Cloud computing applications in agriculture,"

https://www.eescorporation.com/cloud-computing-applications-in-agriculture/

[8] "Cloud manufacturing," Wikipedia, the free encyclopedia

https://en.wikipedia.org/wiki/Cloud_manufacturing

[9] J. S. Saini et al., "[Retracted] cloud computing: Legal issues and provision," *Security and Communication Networks*, vol. 2022,

https://www.hindawi.com/journals/scn/2022/2288961/

[10] "Unlocking the potential of satellite-based cloud computing for space exploration," November 2023,

https://bisresearch.com/insights/unlocking-the-potential-of-satellite-based-cloud-computing-for-space-exploration

[11] S. Malhotra, "Is space cloud the next frontier in cloud computing?" April 2024,

https://www.cio.inc/space-cloud-next-frontier-in-cloud-computing-a-24798

[12] "Why we're using cloud computing for space exploration," August 2020,

https://transmitter.ieee.org/why-were-using-cloud-computing-for-space-exploration/

[13] "Scientists sound the alarm over a growing trash problem in space,"

https://time.com/6262389/space-junk-increasing-problem/

[14] How computing is impacting space exploration," May 2022,

https://voyagerspace.com/insights/how-computing-is-impacting-space-exploration-2/

[15] M. N. O. Sadiku, *Cloud Computing and Its Application*s. Moldova, Europe: Lambert Academic Publishing, 2024.

[16] C. Yang et al., *Cloud Computing in Ocean and Atmospheric Sciences*. Elsevier Science, 2016.

[17] P. Kawade and D. Christodoulopoulou, *Space Cloud Computing*. Analysys Mason, 5th edition, 2024.

CHAPTER 7
INTERNET OF THINGS IN SPACE EXPLORATION

"Just like any company that blissfully ignored the Internet at the turn of the century, the ones that dismiss the Internet of Things risk getting left behind."

—Jared Newman

7.1 INTRODUCTION

The exploration of space has always been a fascinating subject for humans. Space technologies have countless applications on earth and significantly contribute to economic growth, employment, innovation, and societal progress. The spacecrafts with their sensors and communication systems can relay the information to the orbiters, which can further transmit the data to the earth ground stations

Internet of things (IoT) refers to the idea of connecting virtually any electronic device to the Internet and connecting them through the Internet to each other. The rate of connected devices is overwhelmingly increasing and continuously growing with time. With every addition of an IoT object, there is an increase of information in the form of data in the Internet. In various IoT applications, end devices/sensors and controllers are dispersed across a wide geographical area. The advent of IoT has propelled various "smart" or "intelligent" systems across various fields, such as smart homes, smart cities, logistics, manufacturing, agriculture, intelligent transport systems, etc. The IoT is playing a critical role in enabling this new era of space exploration. IoT in space exploration is a critical step in human evolution. It is a key technology for interplanetary exploration, enabling connected devices to communicate and work together. It will open up new possibilities for scientific discovery and unlock business opportunities in industries like telecommunications, agriculture, logistics, and environmental monitoring.

One of the primary ways IoT is used in space exploration is through the deployment of small satellites, which are equipped with IoT sensors that collect data on everything from weather patterns to radiation levels. With IoT, communication can also be established between spacecraft, rovers, and other connected devices, allowing for more efficient coordination and information sharing [1].

The Internet of things (IoT) broadly refers to the ecosystem where billions of interconnected physical objects/devices/end-nodes are equipped with communication, sensing, computing, and actuating capabilities. It is a key technology in space exploration that enables real-time communication, data collection, and safety monitoring. The integration of the IoT in space exploration and colonization represents a major leap forward in technological innovation, allowing for continuous data collection, automated operations, and enhanced scientific discovery [2].

This chapter examines the integration of IoT in space exploration. It begins with providing an overview on IoT. It covers space IoT. It presents some applications of IoT in space exploration. It highlights some benefits and challenges of IoT in space exploration. The last section concludes with comments.

7.2 OVERVIEW ON INTERNET OF THINGS

The Internet began with some military computers in the Pentagon called Arpanet in 1969. It expanded throughout the 1980s as a set of four parallel military networks, each at a different security level. The core technology which gives the Internet its particular characteristics is called Transmission Control Protocol/ Internet Protocol (TCP/IP), which is essentially a set of rules for communication [3].

Internet of Things (IoT) is a worldwide network that connects devices to the Internet and to each other using wireless technology.

IoT is expanding rapidly and it has been estimated that 50 billion devices will be connected to the Internet by 2020. These include smart phones, tablets, desktop computers, autonomous vehicles, refrigerators, toasters, thermostats, cameras, alarm systems, home appliances, insulin pumps, industrial machines, intelligent wheelchairs, wireless sensors, mobile robots, etc. A typical IoT is shown in Figure 7.1 [4].

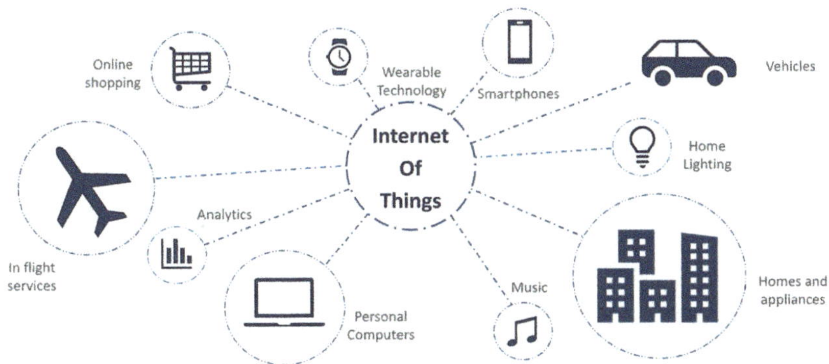

Figure 7.1 A typical IoT [4].

There are four main technologies that enable IoT [5]: (1) Radio-frequency identification (RFID) and near-field communication, (2) Optical tags and quick response codes: This is used for low cost tagging, (3) Bluetooth low energy (BLE), (4) Wireless sensor network: They are usually connected as wireless sensor networks to monitor physical properties in specific environments. These and other technologies associated with IoT are shown in Figure 7.2 [6], while Figure 7.3 illustrates typical IoT applications [7].

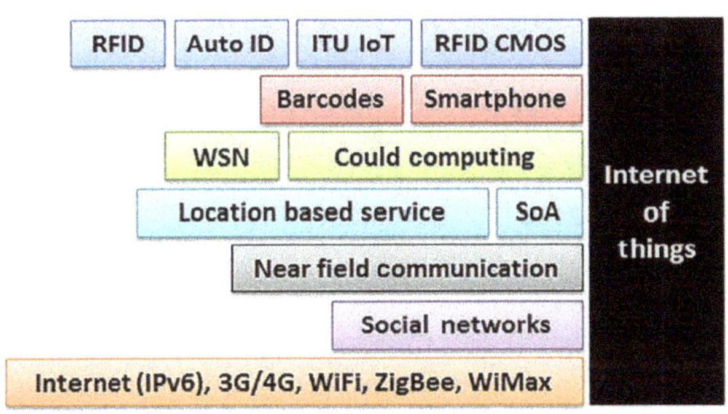

Figure 7.2 Technologies associated with IoT [6].

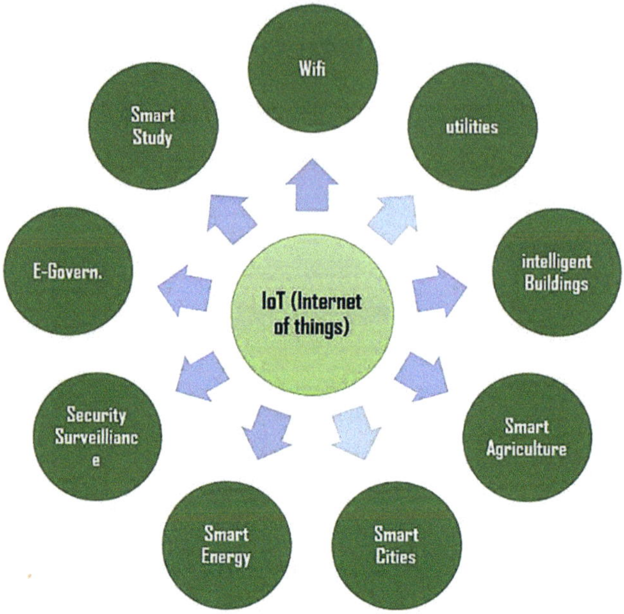

Figure 7.3 Typical IoT applications [7].

IoT technology enables people and objects to interact with each other. It is employed in many areas such as smart transportation, smart cities, smart energy, emergency services, healthcare, data security, industrial control, logistics, retails, structural health, traffic congestion, manufacturing, and waste management. The Internet of things is extensively developed world-wide with a focus on civilian applications such as electric power distribution, intelligent transportation, healthcare, industrial control, precision agriculture, environmental monitoring, etc. Military logistic is perhaps the most important application domain for IoT [8].

7.3 SPACE INTERNET OF THINGS

Space exploration has fascinated humans for centuries. It is no longer improbable and limited to large national government initiatives. It has become an active area of research in academia and industry. There are already many exciting initiatives developed by small startup companies to large private aerospace companies. Elon Musk's SpaceX has also recently become the first private company to launch a spacecraft to the International Space Station

(ISS). Popular media has often discussed the "Mars Race" or "Race to Mars."

Back in 2015, the National Aeronautics and Space Agency (NASA) created a lab dedicated to the study of IoT and space-related applications. NASA set up a lab at the Johnson Space Center and created four teams to research multiple aspects of IoT: security, protocols and monitoring, data analytics, and end user experience. NASA has an extensive network of antennas around the globe — over all seven continents — to receive transmissions from spacecraft. In addition to direct-to-earth communications, many NASA missions rely on relay satellites in order to get their data to the ground. Currently, NASA relies primarily on radio waves for communications, but the agency is developing ways to communicate with infrared lasers. NASA's Laser Communications Relay Demonstration (LCRD) will showcase the benefits of optical communications. Figure 7.4 shows the Laser Communications Relay Demonstration [9].

Figure 7.4 NASA's Laser Communications Relay Demonstration [9].

7.4 APPLICATIONS

Traditional popular space applications include the use of satellites to support telecommunications and global positioning systems (GPS). Key areas of IoT in space include satellite communications, edge computing, Internet of space, aerial network, satellite-terrestrial integrated networks, and space-based 5G networks. Key areas where IoT-integrated space technologies have real-world applications include the following [10-12]:

- *Satellite Communications:* Space communications basically relies on two things: a transmitter and a receiver. Satellites have long been the backbone of global communications, but IoT has taken this to the next level – offering real-time data transfers in areas. Space network consists primarily of the satellite communication network, with the first satellite launched in 1957. Since then, satellite systems have grown exponentially more complex. Large networks of interconnected satellites can provide global coverage for IoT applications like environmental monitoring and asset tracking. For example, Iridium constellation comprises 66 active LEO satellites providing global voice and data coverage in some of the world's most remote regions. Iridium's services connect over a million devices globally, and IoT devices use this network for applications like aircraft tracking, maritime monitoring, and emergency communications. Both NASA and ESA have prepared plans to deploy satellite networks around earth, spanning Mars and the Sun. The networks consist of microwave antenna arrays with miniaturized satellites and lasers pointing through free space. LEO satellites are smaller, cheaper, and easier to produce than their GEO and MEO satellite cousins. It only takes 3 GEO satellites to provide 100% communications coverage at an altitude of 22,000 miles above the Earth's surface. GEO satellites remain visible at all times from a single fixed location on earth. But they are only suited for IoT applications where the content needs to be delivered to the users, without much latency constraints. Satellite-enabled Internet at the LEO level, also known as Satellite Terrestrial Network (STN), is intended to augment the existing terrestrial Internet. Besides the communication subsystem, global positioning system (GPS) is one of the essential but significantly energy-guzzling subsystems in a satellite. Figure 7.5 displays satellite communications [9].

Figure 7.5 Satellite communications [9].

- *Internet of Space:* Space is the next frontier for innovations in IoT, leading to the "Internet of Space." Leading companies around the globe have started seriously considering the concept of the Internet of Space (IoS) or Internet of Space Things (IoST), a cyber-physical system that integrates on-the-ground data and satellite information to enable a wide range of applications. Figure 7.6 shows a typical IoT in Space [13].

Figure 7.6 A typical IoT in Space [13].

It may be regarded as an extension of the Internet of things (IoT) concept into space. The IoST concept was introduced by an innovative group from the Georgia Institute of Technology. It involves a satellite, or a network of them, to address the main challenges in terrestrial IoT deployments – global coverage, scalability, and connectivity. The system is controlled and managed using software-defined networking (SDN) and network function virtualization (NFV). The "things" in IoST specifically refer to a new class of small satellites called CubeSats, indicating that small satellites are the building blocks of Space-IoT. Their role primary revolves around data collection and transmission to and from the earth. They can communicate directly over hundreds of kilometres with various IoT devices on Earth. They enable the realization of a new global cyber-physical system that can be used for a multitude of applications on land, in the air, and in space. Space-IoT relies on a constellation of hundreds of satellites to accomplish global coverage. Security is built into the IoST architecture through the use of different security profiles and delivered as-a-service to protect the availability, integrity, and privacy of all connected resources and information. The IoST can be used for a variety of applications, including monitoring and reconnaissance, in-space backhaul, and cyber-physical integration services. New scientific discoveries will be made possible by the IoST. The system architecture of IoST is shown in Figure 7.7 [14].

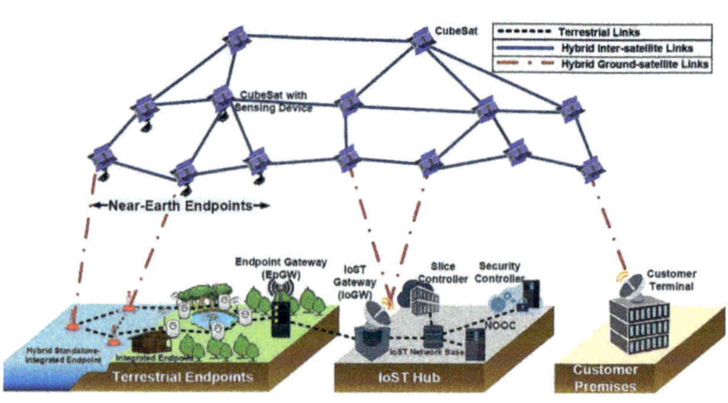

Figure 7.7 The system architecture of IoST [14].

- *Aerial Network:* This has become increasingly popular in the last decade as an intermediate layer between the ground and space networks. Aerial networks are usually served by Unmanned Aerial Vehicles (UAV) and are classified as Low-, Mid- and High-Altitude Platforms, commonly referred to as LAPs, MAPs and HAPs, respectively. UAVs in the form of airships, tethered balloons, and drones, can provide gateway for remote sensor networks, deployed on the ground network. Due to the possibility of limited line-of-sight at higher altitudes and greater distances due to the curvature of the earth and the fact that HAPs might pass beyond the coverage range of ground cellular and IoT networks, connecting UAV networks directly to the satellites is a better alternative.

- *Ground Network:* This is terrestrial (including IoT) networks in the ecosystem. It consist of the satellite ground stations, i.e., the satellite gateways and user terminals, the IoT sensor networks, terrestrial telecommunications networks such as 5G networks, optical networks and wireless local area networks, and IoT gateways. The ground network such as a remote IoT sensor network can be connected to a user satellite terminal. IoT technologies are implemented in ground station operations for efficient monitoring and management of satellite communication. This includes network monitoring, data analytics, security, and automation of ground station functions. Network engineers carefully plan communications between ground stations and missions, ensuring that antennas are ready to receive data as spacecraft pass overhead. Ground station antennas range from the small very high frequency antennas that provide backup communications to the space station to a massive. Figure 7.8 shows two ground antennas located in Fairbanks, Alaska [9].

Figure 7.8 Two ground antennas located in Fairbanks, Alaska [9].

- *Sea Network:* This comprises of sea ships, submarines, underwater sensors, and other maritime communication equipment. One important maritime use case is the tracking of containers and transmitting the sensing data from remote marine locations to the core network. Due to the lack of 4G/5G coverage in the seas and oceans, except ports, space segment is the only feasible option for transmitting and receiving data. Satellite network covers a wide area and provides seamless connectivity to ocean and mountain areas.

- *Robots in Space:* Automation and robotics will be expected to be important for maintaining and aiding activities in space. Different types of large structures can be autonomously assembled in space, in particular, structures that are too large to be launched as one piece into space. Remote control of a large robotic arm for assembly is not feasible given the significant time required for the synchronization of operator commands and actions. Therefore, the autonomous performance of an (even complete) sequence of assembly tasks is required. Robots that assemble structures have been investigated in different projects, though not specifically for space structures. A swarm of robots can construct a structure collaboratively. Apart from large robot arms for assembly or self-assembling parts, there could be robots helping humans within space stations and entertaining human astronauts. Kirobo is Japan's first robot astronaut, which was demonstrated in the International Space Station in 2013. For robots in planets and space, a challenge is how such robots can be empowered by cloud/fog/edge resources. Figure 7.9 shows some space robots [1].

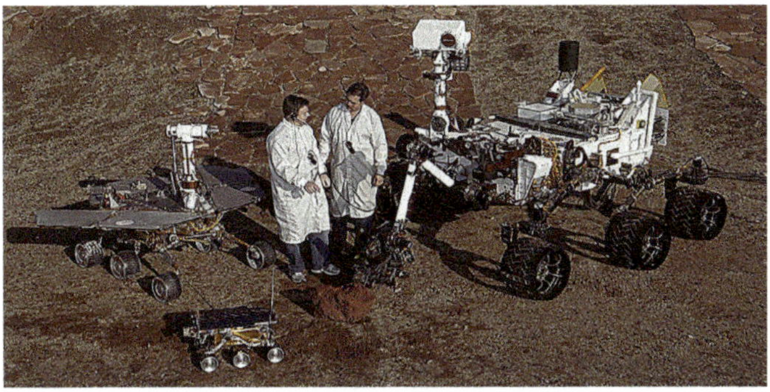

Figure 7.9 Some space robots [1].

- *Space Colonization:* This involves building a colony on another planet. Ideas have been suggested for new horizons of infrastructure development on other planets, including infrastructure for sensor networks, navigation, and data communications that will be needed wherever humans live. Although there is no GPS on other planets, a planetary rover can navigate by computer vision techniques and image comparisons. Sophisticated transportation land and air infrastructure will be needed to accompany human habitats on planets. It has estimated by mathematical simulation that the minimum number of settlers for survival on another planet such as Mars is 110 individuals. Such a population of 110 people will need to be technologically supported. IoT can automate the management of critical systems, such as life support, resource utilization, and environmental control. This could help create self-sustaining colonies on other planets.

7.5 BENEFITS

Integrating IoT technology into space exploration signifies a significant step towards more advanced and comprehensive space missions, opening up new possibilities for scientific discovery and exploration beyond Earth. The integration is unlocking numerous benefits and opportunities for related industries to innovate further. Whether it is CubeSats or LEO satellites, the growing availability of commercially available space parts is allowing more companies to get into the space business. Existing satellites are generally equipped with onboard health monitoring units as a part of the subsystems. Other benefits include the following [11]:

- *Data Collection:* IoT devices on satellites collect vast amounts of data, from climate monitoring to crop health analysis, providing valuable insights that can lead to better resource management on earth. IoT sensors on spacecraft and rovers can continuously monitor various parameters like temperature, pressure, radiation levels, and geological conditions, sending data back to earth for immediate analysis. LEO satellites are often used to collect data in space for applications such as weather forecasting environment monitoring and target surveillance. Once collected, these data need to be downloaded to ground servers via earth Stations.
- *Mission Efficiency:* IoT networks facilitate communication between multiple spacecraft and ground stations, enabling better coordination and decision-making during space missions. Real-time monitoring and predictive maintenance systems enabled by IoT ensure that space missions are more efficient and less prone to failure. This reduces costs and increases the chances of mission success.
- *Global Connectivity:* Satellite communication can be used to provides backhaul connectivity for different types of wireless base stations. Miniaturized satellites, known as CubeSats, has emerged as a viable solution towards establishing global connectivity at low costs. IoT-driven satellite networks provide global connectivity, which is essential for industries like logistics, environmental monitoring, and emergency response. They ensure that no part of the world is out of reach. The Starlink satellites supported by SpaceX aim to provide rural areas with high-speed broadband connectivity previously only available in urban areas.
- *Scientific Discovery:* IoT technologies enable continuous data collection and monitoring in space, leading to new scientific discoveries about our universe and planetary geology, space weather, and even potential life forms. The vast amount of data collected through IoT sensors can lead to new insights into planetary geology, space weather, and potentially even

the existence of extraterrestrial life.

- *Autonomous Operations:* By utilizing IoT connectivity, robots and automated vehicles can perform complex tasks in space with minimal human intervention, like repairs on the International Space Station or exploration on other planets. IoT will enable the development of robots and automated vehicles that can perform complex tasks in space.
- *Space Debris:* IoT is critical for tracking space debris and managing space traffic. With the increasing number of satellites and spacecraft, maintaining situational awareness is vital to avoiding collisions and ensuring safe operations in space. The US Space Surveillance Network operates the Space Fence system, which tracks space debris and satellites in real-time.
- *Colonizing Planets:* As humanity looks to colonize other planets, IoT will ensure that life support systems, habitats, and infrastructure are functional, sustainable, and safe, making long-term human presence in space feasible.
- *Safety:* IoT is also being used to improve the safety and sustainability of space exploration. It can monitor astronaut health and safety, track spacecraft movement, and recycle waste. Connected devices can monitor the health and safety of astronauts, track the movement of spacecraft, and even recycle waste into usable resources. This technology can help extend the lifespan of exploration missions and make space exploration more sustainable in the long term.
- *Data Transmission:* IoT can transmit data between sensors and devices through space, even in areas with poor network coverage on earth. Satellites can act as a transmission bridge, enabling data transfer at any time.
- *Environment Sensing:* This involves sensing the environment using satellites. Project SatSense is focused on using satellite telemetry to develop new techniques for environmental monitoring. We believe this will lead to gaining accurate insights about today's ecosystem at large-scale including

forestry, agriculture, oil & gas, etc. Figure 7.10 illustrates environmental sensing using satellites [15].

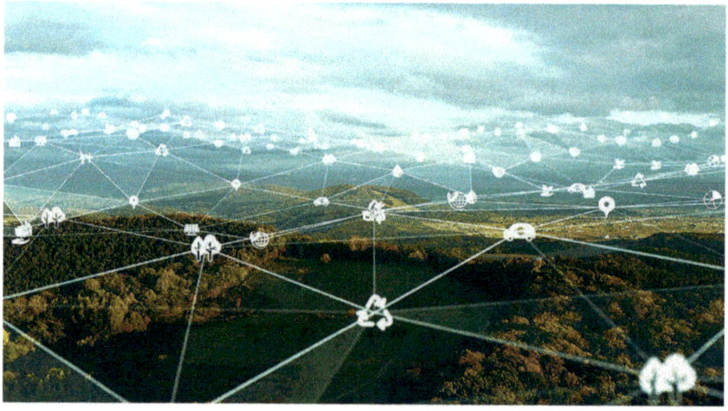

Figure 7.10 Environmental sensing using satellites [15].

7.6 CHALLENGES

To truly jumpstart space markets, commercial off-the-shelf costs will have to come down for high-quality components. It is essential to address the key challenges, such as scalability, ubiquitous global coverage, and real-time connectivity, that arise due to this immense growth of IoT devices in space exploration. One of the main challenges in inter-planetary communications is free space loss. While the potential of IoT in space is vast, several challenges must be addressed [11,16]:

- *Robust Communication:* Space is a challenging environment for IoT systems due to high radiation levels and potential signal interference from cosmic sources. Developing radiation-hardened components and robust communication protocols is essential to ensure the reliability of IoT systems in space. IoT devices must be designed to withstand extreme temperatures, radiation, and vacuum conditions.

- *Interference:* As communications transmissions travel over long distances or through the atmosphere, the quality of their data can deteriorate, garbling the message. Radiation from other missions, the Sun, or other celestial bodies can also interfere with the quality of transmissions. Radiation from the Sun is one potential source of interference for space communications transmissions.

- *Space Pollution:* The increasing number of satellites and IoT devices in space contributes to the growing problem of space debris. Companies like Astroscale are developing technologies to remove defunct satellites from orbit.
- *Power Constraints:* Spacecraft often have limited power sources, requiring energy-efficient IoT devices. Resources are severely constrained in terms of size, mass, and available power. Due to the sheer distance of other planets from the earth, the power required for sending data is large. While big satellites typically do not have any constraints on energy consumption for GPS subsystem, such is not the case in miniaturized satellites.
- *Communication Latency:* Most IoT applications are bounded by delay or latency requirements. Communications do not take place instantaneously. They are bound by a universal speed limit: the speed of light, about 186,000 miles per second. IoT applications can be mainly categorized into delay-tolerant applications and delay-sensitive applications. Signal delay in deep space can pose challenges for real-time data transmission. For GEO satellites, the minimum propagation delay adds up to half a second, which is not suitable for time-critical IoT applications, such as sensor networks for autonomous driving. The ability to provide broadband access depends on the low-latency, high-speed communications provided by low earth orbit (LEO) satellite constellations.
- *Cost:* One challenge with the space industry is the cost of developing and launching satellites, which makes access to space expensive. Costs are based on the data rate because IoT is still in the development stage. The considerable reduction in launch costs and the ability to reuse rockets will help drive down costs in space travel.
- *Space Debris:* These are problematic if space is to be further populated. The challenge of tracking people and things in space will be enormous, not only in detecting and

tracking space debris, but also in tracking people and things. Cooperation among different parties is required to be able to adequately track objects in space to ensure the safety of space flights. Space situational awareness is required if collisions are to be avoided and the space within the earth's orbit is to be best utilized.

- *Security:* This has been a contentious topic of conversation. We are still not sure how companies will deal with the massive amounts of data IoT will generate. As more devices become connected, the risk of cyber-attacks, data breaches, and unauthorized access increases. Ensuring the security and privacy of user data is critical to maintaining user trust and avoiding potential financial and reputational damage.
- *Standardization:* With numerous manufacturers and vendors in the market, there is a lack of consistent standards and protocols for communication between devices. This can limit the functionality and efficiency of IoT systems, as devices may not work seamlessly with each other.
- *Regulation:* IoT devices and systems are subject to various legal and regulatory requirements, including data protection laws, consumer protection laws, and industry-specific regulations. Space-based IoT systems require international collaboration, agreements, and regulation. International regulations are needed to manage space traffic and ensure the sustainability of space exploration. Organizations like the International Telecommunication Union (ITU) play a critical role in managing the allocation of radio frequencies and ensuring that space-based IoT systems do not interfere with one another.
- *Data Management:* The IoT industry generates vast amounts of data, which can be challenging to manage, store, and analyze effectively. Companies must develop robust systems for handling this data to derive valuable insights and make informed decisions.
- *Scalability:* As the number of IoT devices grows, it becomes increasingly difficult to manage and maintain them. Ensuring

that networks, systems, and infrastructures can scale to accommodate growth is a significant challenge.

7.7 CONCLUSION

The space industry is projected to experience skyrocketing growth. When it comes to applications in space, IoT currently has more conceptual uses than physical applications. IoT enables remote monitoring in space missions by using sensors and communication networks to collect and transmit data from spacecraft and equipment continuously. The IoT has the potential to change our space exploration immensely, presenting new opportunities as well as new challenges. The enthusiasm for space exploration is driving innovation across the IoT. As interplanetary exploration continues to expand, the importance of IoT in this field will only continue to grow and enable real-time communication, data collection, safety monitoring, and sustainability. More information about IoT in space can be found in the books in [17,18].

REFERENCE

[1] "IoT in space: Enabling interplanetary exploration with connected devices," October 2024,

https://ijcsmc.com/index.php/iot-in-space-enabling-interplanetary-exploration-with-connected-devices/

[2] M. N. O. Sadiku, U. C. Chukwu, and J. O. Sadiku, "Internet of things in space exploration," *Innovative Mult-disciplinary Journal of Applied Technology*, vol. 2, no. 10, 2024, pp. 240-252.

[3] M. Townes, "The spread of TCP/IP: How the Internet became the Internet,"

Millennium: Journal of International Studies, vol. 41, no. 1, 2012, pp. 43 –64.

[4] "IoT for developers: Internet of things history, learning," May 2020,

https://programmerblog.net/iot-for-developers/

[5] M. N. O. Sadiku, S.M. Musa and S. R. Nelatury, "Internet of things: An Introduction," *International Journal of Engineering Research and Advanced Technology*, vol. 2, no.3, March 2016, pp. 39-43.

[6] S. S. Tabrizi and D. Ibrahim, "Security of the Internet of things: an overview,"

Proceedings of the 2016 International Conference on Communication and Information Systems, December 2016, pp. 146-150.

[7] "Internet of things (IoT),"

https://www.geeksforgeeks.org/internet-things-iot-2/

[8] L. Yushi, J. Fei, and Y. Hui, "Study on application modes of military Internet of things (MIOT)," *IEEE International Conference on Computer Science and Automation Engineering*, May 2012, pp. 630-634.

[9] K. Schauer, "Space communications: 7 Things you need to know," October 2020,

https://www.nasa.gov/missions/tech-demonstration/space-communications-7-things-you-need-to-know/

[10] J. Kua et al., "Internet of things in space: A review of opportunities and challenges from satellite-aided computing to digitally-enhanced space living," Sensors, vol. 21, no. 23, December 2021,

https://www.mdpi.com/1424-8220/21/23/8117

[11] "IoT in space exploration: Significant step for innovation, giant leap for civilization,"

https://www.ignitec.com/insights/iot-in-space-exploration-significant-step-for-innovation-giant-leap-for-civilisation/#:~:text=IoT%20in%20space%20exploration%20is,driving%20economic%20growth%20and%20innovation.

[12] S. Narayana, "Space Internet of Things (Space-IoT),"

https://research.tudelft.nl/en/publications/space-internet-of-things-space-iot

[13] K. Mccann, "What innovations will IoT in space bring? | Interview with Air Direct Solutions' Paul Kostek," June 2023,

https://www.iotinsider.com/industries/industrial/what-innovations-will-iot-in-space-bring-interview-with-air-direct-solutions-paul-kostek/

[14] "Internet of Space Things," August 2019,

https://licensing.research.gatech.edu/technology/internet-space-things

[15] "6G | Space: Internet of things (IoT)," July 2023,

https://www.microsoft.com/en-us/research/articles/6g-space-internet-of-things-iot/

[16] "How can space technology help the Internet of things?"

https://newspaceeconomy.ca/2023/04/02/how-can-space-technology-help-the-internet-of-things/

[17] M. N. O. Sadiku, *Internet of Things and Its Applications*. Moldova, Europe: Lambert Academic Publishing, 2024.

[18] F. Xhafa et al. (eds.), *IoT and Spacecraft Informatics*. Elsevier Science, 2022.

CHAPTER 8
BLOCKCHAIN IN SPACE EXPLORATION

"The blockchain is a distributed network that solves all the problems that we have of finance, but more broadly, it's like a philosophy. It's a way of life."

—Mike Cernovich

8.1 INTRODUCTION

Recently, modern technologies have aided business titans in deciding to use digital currency rather than conventional trade in order to avoid fraud. Blockchain enables the existence of Bitcoin and many other cryptocurrencies. A cryptocurrency refers to a digital asset that works as a medium of exchange between various business organizations. Figure 8.1 shows the symbol of blockchain [1].

Figure 8.1 The symbol of blockchain [1].

The evolution of human interest in space has been driven by the desire to explore the unknown, pushing the boundaries of knowledge with the urge to discover new worlds. Starting with the curiosity about the sky, moon, sun, and other stars visible to human eye, we have seen the first human to walk on the moon, as typically shown in Figure 8.2 [2].

Figure 8.2 An astronaut [2].

This can be referred to as moving from Space1.0 to Space 2.0. The permanent presence in space with the International Space Station was the benchmark for the evolution from Space 2.0 to Space 3.0. Today we are in the middle of a full-scale evolution of the space sector into a new era, namely Space 4.0 [3]. Blockchain technology has transcended its financial origins to find applications in various industries, including space industry. It is an ideal candidate for revolutionizing the space exploration. It can enhance the security of space infrastructure.

Blockchain, the backbone of cryptocurrencies like Bitcoin, is a distributed ledger technology, in which blocks of data are chained together by cryptographic hashing. It operates on a decentralized network of computers, ensuring transparency, security, and immutability of data. Integrating blockchain in space operations introduces decentralized systems, reducing reliance on a central authority. Blockchain can serve as one of the potential technology solutions to help remediate some of the challenges confronting space exploration. In this paper, we embark on a journey to unravel the implications of marrying blockchain technology with space concepts.

Blockchain, the backbone of cryptocurrencies like Bitcoin, is a distributed ledger technology, in which blocks of data are chained together by cryptographic hashing. It operates on a decentralized network of computers, ensuring transparency, security, and immutability of data. Integrating blockchain in space operations introduces decentralized systems, reducing reliance on a central authority. Blockchain can serve as one of the potential technology solutions to help remediate some of the challenges confronting space exploration [4].

In this chapter, we embark on a journey to unravel the implications of marrying blockchain technology with space concepts. It begins with describing what blockchain is all about. It covers blockchain in space exploration. It presents some applications of blockchain in space exploration. It highlights some benefits and challenges of blockchain in space exploration. The last section concludes with comments.

8.2 WHAT IS BLOCKCHAIN?

Blockchain, a type of distributed digital ledger technology (DLT), is a relatively new and exciting way of recording transactions in the digital age. It is a decentralized and distributed digital ledger technology that securely records and verifies transactions across multiple computers or nodes in a network. Basically, it is a chain of blocks in which each block contains a list of transactions. The blockchain technology was created as the foundational basis for Bitcoin – a digital currency in which secure peer-to-peer transactions occur over the Internet. It is expected that the spending on blockchain solutions worldwide would grow from 4.5 billion USD (2020) to an estimated value of 19 billion USD by 2024 [5].

Originally developed as the accounting method for the virtual currency Bitcoin, Blockchains are appearing in a variety of commercial applications today. Blockchain technology is a type of distributed digital ledger that uses encryption to make entries permanent and tamper-proof and can be programmed to record financial transactions. It is used for secure transfer of money, assets, and information via a computer network such as the Internet without requiring a third-party intermediary. It is now being adopted across financial and non-financial sectors. As a catalyst for change, the Blockchain technology is going to change the business world and financial matters in major ways.

The first Blockchain was conceived in 2008 by an anonymous person or group known as Satoshi Nakamoto, who published a white paper introducing the concept of a peer-to-peer electronic cash system he called Bitcoin [6,7]. Bitcoin and Ethereum are the first two mainstream blockchains. Other modern blockchains include Namecoin, Peercoin, Ether, and Litecoin. Figure 8.3 shows different components of blockchain [8].

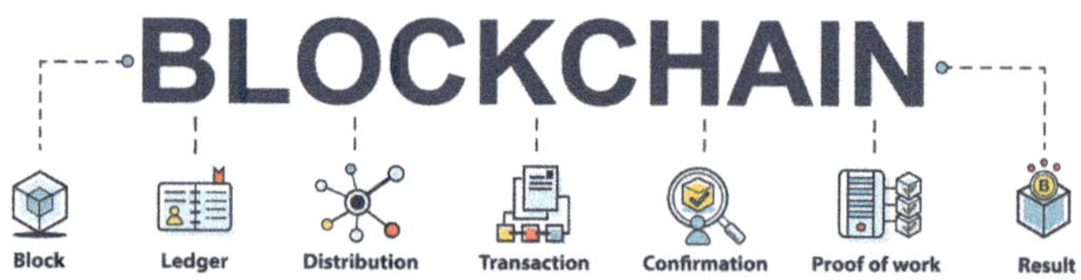

Figure 8.3 Different components of Blockchain [8].

Blockchain combines existing technologies such as distributed digital ledgers, encryption, immutable records management, asset tokenization and decentralized governance to capture and record information that participants in a network need to interact and transact. As illustrated in Figure 8.4, a complete blockchain incorporates all the following five elements [9]:

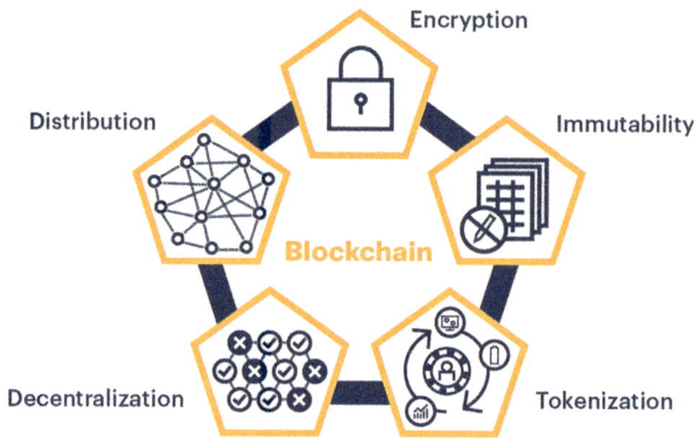

Figure 8.4 Five key elements of Blockchain [9].

- *Distribution:* Digital assets are distributed, not copied or transferred. A protocol establishes a set of rules in the form of distributed mathematical computations that ensures the integrity of the data exchanged among a large number of computing devises without going though a trusted third party. A centralized architecture presents several issues including a single point of failure and problems of scalability.

- *Encryption:* BC uses technologies such as public and private keys to record data securely and semi-anonymously. Completed transactions are cryptographically signed, time-stamped, and sequentially added to the ledger.
- *Immutability:* The Blockchain was designed so these transactions are immutable, i.e. they cannot be deleted. No entity can modify the transaction records. Thus, Blockchains are secure and meddle-free by design. Data can be distributed, but not copied.
- *Tokenization:* Value is exchanged in the form of tokens, which can represent a wide variety of asset types, including monetary assets, units of data or user identities.
- *Decentralization:* No single entity controls a majority of the nodes or dictates the rules. A consensus mechanism verifies and approves transactions, eliminating the need for a central intermediary to govern the network.

Bitcoin and its underlying blockchain technology increasingly impact all facets of society. Bitcoin's status as digital gold is merely the tip of this technology. Figure 8.5 shows Bitcoin [10]. Although blockchain technology will for all time be associated with Bitcoin due to their common genesis, it has broader applications. Cryptocurrency will increasingly become a factor in family law issues as well.

Figure 8.5 Bitcoin [10].

8.3 BLOCKCHAIN IN SPACE EXPLORATION

Space exploration has come a long way since the first moon landing in 1969. The space industry has been experiencing an unprecedented surge in recent years and is expected to continue to grow. It is currently undergoing rapid evolution as private companies or other entities transform the sector and spearhead innovation in space technologies. Figure 8.6 shows a typical space exploration [2].

Figure 8.6 A typical space exploration [2].

Spacecraft and satellites require a secure data communication network. Satellites present an interesting opportunity on which to build a blockchain network that would enable a decentralized flow of data between machines. A conceptual view of a blockchain network is presented in Figure 8.7 [11].

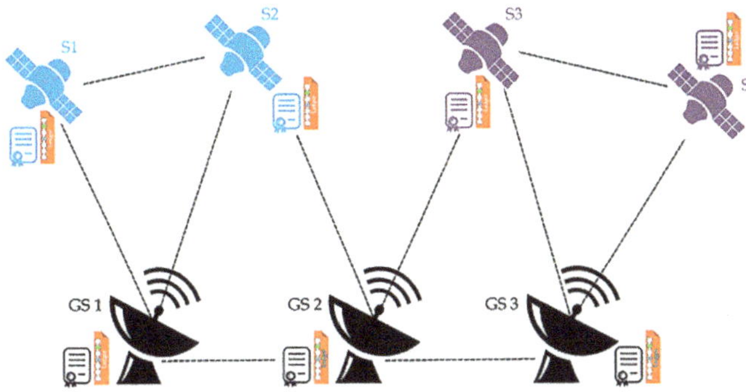

Figure 8.7 A conceptual view of a blockchain network [11].

The rise of blockchain technology has transformed how we conduct transactions and share information online. National Institute of Standards and Technology Blockchain, or the more general term, distributed ledger technology (DLT), is gaining traction across various industries outside the space sector. NASA has embraced blockchain technology in various projects, including secure communication protocols and tracking the procurement of space equipment. It is looking to use smart contracts on the Ethereum blockchain in the agency's SensorWeb program. The main objective of the program is to create an interoperable environment for a diverse set of satellite sensors via the use of software and the Internet.

8.4 APPLICATIONS OF BLOCK CHAIN IN SPACE EXPLORATION

Blockchain technology is so new that it seem difficult to see direct applications in space exploration. One example of blockchain in space is AstroChain, which provides a platform for space enthusiasts to share astronomy content and images. While the most popular use of cryptocurrency has been in the financial sector, it has been applied for space. Other applications of blockchain in space include the following [12-15]:

- *Space Economy:* The space economy has evolved significantly from its initial focus on government-funded projects to a vibrant ecosystem involving private companies, international collaborations, and a surge in commercial space activities.

Blockchain technology has the potential to revolutionize the space economy by introducing greater transparency, efficiency, and accessibility to space-related activities, fostering a new era of collaboration and innovation in the cosmic frontier. It has the potential to significantly disrupt and enhance the space economy by providing a secure, transparent, and decentralized system for managing assets, tracking data, and facilitating transactions across various space operations.

- *Space Debris:* This is a growing challenge that threatens the future of space exploration and satellite operations. As the number of satellites and space missions increases, so does the accumulation of debris, posing risks to spacecraft, satellites, and even the International Space Station (ISS). Addressing this issue requires innovative solutions, and blockchain technology emerges as a promising tool to enhance space debris management and promote sustainable space exploration. Blockchain can enhance space debris tracking efforts by providing a means for real-time open source data sharing on space debris, while developing new incentive mechanisms to promote sustainable space exploration. One of the primary applications of blockchain in space debris management is the creation of a decentralized database for tracking and identifying debris. Blockchain's smart contracts offer an automated mechanism for executing agreements related to space debris removal. The integration of blockchain technology into space debris management represents a significant step toward ensuring the long-term sustainability of space exploration. Figure 8.8 shows some space debris [12].

Figure 8.8 Some space debris [12].

- *Supply Chain Management:* Another area where blockchain is revolutionizing space exploration is supply chain management. Managing the supply chain in space is a complex process. Blockchain technology can streamline the supply chain by creating a transparent and immutable record of each transaction. This will help to reduce errors, delays, and inefficiencies, making the supply chain more cost-effective and reliable. Blockchain technology can support the development of space-based supply chain management systems. Space infrastructure requires a constant flow of resources and materials, and blockchain can provide a secure, transparent platform for tracking the movement of goods and ensuring supply chain efficiency. Blockchain can enhance the supply chain management of recycling materials, ensuring the traceability and authenticity of recycled components. This not only supports the circular economy within the aerospace industry but also promotes the responsible use of resources in space manufacturing. Supply chain logistics are challenging when dealing with space, given the vast distances and communication lags involved.

- *Earth Observation:* Another growing industry that has significant growth potential involves Earth Observation. This is an evolving sector that involves using remote sensing technologies to collect information on the planet's physical, chemical, and biological systems and to monitor land, water (i.e., seas, rivers, lakes), as well as the atmosphere. Once analyzed, information collected through earth observation can be assimilated into detailed models to produce information and intelligence Space-based geographical applications and earth observation data are highly multidimensional with coordinates, timestamps, different data from different sensors, data from different suppliers with different revenue schemes. Eventually everyone can access space data, not only governments, intelligence agencies or big corporates. Blockchain can assure satellite data is accessed by everyone in a trusted manner as the authenticity and the integrity of data will be assured in a distributed ownership infrastructure.

- *Satellite Communication:* Satellite communication is vital to the global telecommunications network, enabling essential services - calls, internet access, TV broadcasting, and radio channels. In a standard satellite communication link, a signal is transmitted or uplinked from our planet to a designated place - a station. Traditional satellite networks rely on centralized infrastructure and control systems, leading to single points of failure and vulnerabilities. A typical satellite communication system is shown in Figure 8.9 [13].

Figure 8.9 A typical satellite communication system [13].

Blockchain technology can be used in satellite communication to enhance data security, transparency, and integrity by creating a decentralized and immutable ledger for storing satellite data. Decentralized satellite networks enhance resilience, security, and efficiency, enabling global connectivity, remote sensing, and satellite-based services. The integration of blockchain technology with satellite communication heralds a new era of possibilities. Blockchain technology gives the possibility to establish a decentralized communication network that boasts enhanced protection when dealing with cyber risks. It could track a satellite's movement, sharing data with all suppliers, and can enforce rules such as any changes made to the satellite require the consensus of the team. Figure 8.10 shows the integration of blockchain technology and satellite communications [14].

Figure 8.10 Integration of blockchain technology and satellite communications [14].

- *Space Colonization:* In recent years, humanity's gaze has shifted from our earthly confines to the vast expanse of space. This new era of exploration is driven by a collective ambition to transcend our planetary boundaries and establish a human presence among the stars. The ultimate target in this space race is Mars — a planet that has long captivated our imagination. With robotics like the Mars Rover already providing invaluable data from the Martian surface, the dream of human colonization feels closer than ever. Blockchain technology can potentially propel humanity towards Mars colonization by providing a transparent, secure, and decentralized system for managing crucial aspects of a Martian colony, including resource allocation, property ownership, governance, and even the tracking of vital supplies. It can play a significant role in enabling a more efficient, transparent, and equitable Martian colony by managing resources, property ownership, governance, and economic transactions in a decentralized manner. Companies like SpaceX, Blue Origin, and NASA have been at the forefront of this cosmic endeavor, pushing the limits of technology and human ingenuity to realize this celestial ambition. MarsVerse is the world's first metaverse ecosystem with Web3 interstellar colonization at its core. It recruits Mars colonization volunteers globally through Avatars to help humanity officially enter the universe.

8.5 BENEFITS

Integrating blockchain technology in space operations introduces decentralized systems, reducing reliance on a central authority. Blockchain can be used in space exploration to enhance security, transparency, and trust in many ways. Establishing trust in decentralized space systems involves demonstrating the reliability, security, and efficiency of blockchain applications. Space-based blockchain can provide a fully decentralized and sovereignty agnostic infrastructure. Blockchain technology presents a unique opportunity to enhance space communications, offering a more secure, efficient, and accessible infrastructure. Other benefits include [10,16,17]:

- *Security:* This is paramount in space missions. Blockchain's cryptographic algorithms provide robust security, protecting sensitive data from potential cyber threats. As space missions involve confidential information and critical communication, blockchain acts as a shield against unauthorized access and tampering. Blockchain's cryptographic principles offer a robust solution to enhance the security of satellite communication.
- *Improved Transparency:* Transparency is a cornerstone in space-related transactions, especially in collaborations involving multiple entities. Blockchain's transparent and traceable nature ensures that every transaction is recorded and accessible, fostering trust among stakeholders.
- *Smart Contracts:* Smart contracts, a key feature of blockchain, can automate and enforce agreements between satellite operators, ground stations, and other service providers. They are self-executing contracts coded into the blockchain that automatically perform actions once predefined conditions are met. They streamline processes in the space economy. They automate data transactions and enforce data sharing agreements. Automating tasks like satellite launches, resource allocation, and data sharing through smart contracts reduces operational costs and minimizes the risk of human error. A smart contract is constrained by the boundaries of the database in which the smart contract resides.
- *Space Tourism:* The burgeoning space tourism industry can leverage blockchain for secure transactions, identity verification, and customer data protection. This synergy enhances the overall experience for space tourists while addressing privacy concerns.
- *Decentralized System:* Current data management systems rely heavily on centralized servers on earth, leading to risks of single points of failure and inefficiencies in data handling. Blockchain's decentralized nature offers an innovative solution to these challenges. Integrating blockchain in space operations introduces decentralized systems, reducing

reliance on a central authority. This decentralization enhances the resilience of space networks, reducing the risk of single points of failure and ensuring uninterrupted communication and data transfer.

- *Interplanetary Communication:* As humanity prepares for interplanetary missions, the need for secure and reliable communication will become more pressing. Blockchain could help create an "interplanetary Internet," enabling decentralized, secure, and tamper-proof communication between Earth, Mars, the Moon, and beyond.

- *Securing Communications:* The vast distances involved in space exploration require secure and reliable communication systems. Blockchain technology can enhance the security and authenticity of communication protocols. By creating a blockchain-based authentication system, NASA can ensure that messages sent between earth and spacecraft remain confidential and tamper-proof.

- *Satellite Coordination:* In an era where an increasing number of satellites are being launched into orbit, space traffic management is becoming a pressing concern. Blockchain can be employed to create a decentralized and transparent system for coordinating satellite movements. With real-time updates on the blockchain, satellite operators can make informed decisions to safeguard valuable assets in space.

8.6 CHALLENGES

The space industry faces numerous challenges ranging from data management to resource allocation. Foundational technologies like blockchain must first overcome technological, organizational, regulatory, and political barriers. Other challenges include [9,15]:

- *Security:* Although blockchain enhances security, implementing it in space introduces unique challenges. Robust solutions involving space-hardened hardware and continuous monitoring are crucial to mitigate these challenges.

- *Regulation:* The integration of blockchain in space operations raises regulatory concerns. Establishing clear regulatory frameworks to govern space-based blockchain activities is essential. Since governments play a pivotal role in shaping the future of blockchain in space, collaborative efforts between governments and space agencies are essential to create a regulatory framework. The development of standards and regulations for blockchain use in the space industry is still evolving.

- *Collaboration:* Collaborations between technology and aerospace industries foster innovation. Partnerships can lead to the development of integrated solutions that address the unique challenges of space while advancing blockchain technology. Blockchain technology enables collaboration without compromising sensitive information, enhancing joint efforts in debris monitoring, threat assessment, and the development of comprehensive mitigation strategies.
- *Sustainability:* As space activities increase, environmental sustainability becomes paramount. Implementing eco-friendly practices in space exploration, supported by blockchain technology, can minimize the ecological footprint of space missions. The potential for recycling materials recovered from space debris presents an opportunity for sustainability in space operations.
- *Harsh Environments:* Harsh space conditions can impact data integrity. Employing advanced encryption methods and developing space-hardened blockchain solutions are imperative to safeguard the reliability of data in space-based applications.
- *Complexity:* Integrating blockchain technology with existing space systems requires sophisticated development and adaptation. Blockchain for space systems may offer compelling advantages by reducing complexity across a range of business, operational, and security applications. NASA as well as many other businesses may still need to integrate with older technology. Another layer of complexity is added when we take into account the possibility of fractional ownership.
- *Memory Constraints:* One of the challenges in working with

devices is the memory constraints. This is amplified on a satellite because much of the memory is reserved as a backup for the satellite itself. So we need to creative to be able to generate the private keys for new accounts.

- *Cost:* One of the main challenges that have always slowed down any attempt by humanity to research the cosmos involves acquiring the funds necessary to leave earth's orbit. Recently, space was only allowed to governments or space agencies due to extremely high costs and operational risks which made space exploration inaccessible to private companies. Thanks to the advanced satellite, launch, or other space technologies, we now witness the emergence of other players in space exploration, colonization, and commercialization. Recently, SpaceX launches Starlink LEO satellites, as an example of such a new entrepreneurial spirit that is rapidly shaping a new space economy.
- *Data Management:* Space missions generate vast amounts of data, including telemetry data, sensor readings, and imagery, which must be managed, analyzed, and shared efficiently. For example, NASA's Mars Rover missions produce hundreds of gigabytes of data that are transmitted back to earth. This data needs to be securely stored, shared, and analyzed, often in real time. Blockchain-based solutions provide secure, transparent, and immutable data management for space missions, ensuring data integrity, provenance, and traceability.
- *Scalability:* Current blockchain technology faces limitations in handling large volumes of data transactions, a critical factor for space communication. Integrating blockchain with large-scale satellite networks may pose scalability challenges due to the computational demands of consensus mechanisms.
- *Latency:* The inherent latency associated with blockchain transactions could pose a challenge for real-time space communication applications. Depending on the network architecture, latency in satellite communication could impact blockchain transaction speeds

8.7 CONCLUSION

Blockchain is an emerging technology that is on the verge of transforming many industries, and space exploration is no exception. The growing interest in blockchain development for space applications indicates that it could very well be the future of space missions. As humanity ventures further into the cosmos, blockchain may become as integral to space exploration as rockets and rovers. As the industry continues to grow, blockchain technology is poised to play an increasingly important role in shaping the space industry's future. As blockchain moves into outer space, its ability to tokenize spacecraft and payloads is key to its success. Its ability to provide decentralized, secure, and efficient data management solutions is a game-changer for future space missions. More information on the integration of blockchain technology into space exploration is available from the books in [18,19].

REFERENCE

[1] "Blockchain meets the oil & gas industry," February 2018,

https://executiveacademy.at/en/news/detail/blockchain-meets-the-oil-gas-industry/

[2] "The role of Web3 in space exploration: Decentralized solutions for aerospace challenges," February 2024,

https://www.linkedin.com/pulse/role-web3-space-exploration-decentralized-solutions-aerospace-gbktf#:~:text=Web3%20enables%20the%20creation%20of%20collaborative%20space%20research%20platforms%2C%20where,contribute%20to%20space%20exploration%20projects.

[3] O. Gurhan, "Blockchain and the new space age: An inevitable convergence of Web 3.0 and Space 4.0?" October 2023,

https://www.aviatelabs.co/post/blockchain-and-the-new-space-age

[4] M. N. O. Sadiku, P. A. Adekunte, and J. O. Sadiku, "Blockchain in space exploration," *International Journal of Trend in Scientific Research and Development*, vol. 8, no. 5, September-October 2024, pp. 1151-1160.

[5] C. M. M. Kotteti and M. N. O. Sadiku, "Blockchain technology," International Journal of Trend in Research and Development, vol. 10, no. 3, May-June 2023, pp. 274-276.

[6] "Blockchain," *Wikipedia*, the free encyclopedia

https://en.wikipedia.org/wiki/Blockchain

[7] S. Nakamoto, "Bitcoin: A peer-to-peer electronic cash system,"

https://bitcoin.org/bitcoin.pdf

[8] "The beginning of a new era in technology: Blockchain traceability,"

https://www.visiott.com/blog/blockchain-traceability/#:~:text=The%20Beginning%20of%20a%20New,money%20without%20a%20central%20bank.

[9] "The CIO's guide to blockchain,"

https://www.gartner.com/smarterwithgartner/the-cios-guide-to-blockchain#:~:text=True%20blockchain%20has%20five%20elements,%2C%20immutability%2C%20tokenization%20and%20decentralization.

[10] "Blockchain and space exploration: Is decentralized data the future of space missions?" October 2024,

https://medium.com/coinmonks/is-decentralized-data-the-future-of-space-missions-646173d1aeec

[11] R. Mital et al., "Blockchain application within a multi-sensor satellite architecture,"

https://ntrs.nasa.gov/api/citations/20180006549/downloads/20180006549.pdf

[12] "Leveraging blockchain for space debris management: A path to sustainable space exploration," April 2024,

https://decentcybersecurity.eu/leveraging-blockchain-for-space-debris-management-a-path-to-sustainable-space-exploration/

[13] "Exploring blockchain in the African space program," March 2024,

https://medium.com/@africablockchaininstitute/exploring-blockchain-in-the-african-space-program-694f763d9bb9

[14] "The use of blockchain technology in satellite communication," March 2024,

https://simpleswap.io/blog/the-use-of-blockchain-technology-in-satellite-communication

[15] "How can blockchain technology propel humanity towards mars colonization?" July 2024,

https://medium.com/@marsverse/how-can-blockchain-technology-propel-humanity-towards-mars-colonization-

3d53a9388e39

[16] S. H. Abbas, “Blockchain and space economy: Leveraging technology beyond earth,”

https://medium.com/@zee.associates001/blockchain-and-space-economy-leveraging-technology-beyond-earth-b773931b784b#:~:text=Integrating%20blockchain%20in%20space%20operations,uninterrupted%20communication%20and%20data%20transfer.

[17] “How blockchain is revolutionizing space exploration?” September 2023,

https://www.altcoinbuzz.io/bitcoin-and-crypto-guide/how-is-blockchain-revolutionizing-space-exploration/

[18] M. N. O. Sadiku, *Blockchain Technology and Its Applications*. Moldova, Europe: Lambert Academic Publishing, 2023.

[19] S. Williams and G. Pullen, *Blockchain & the Space Economy*. Primedia eLaunch LLC, 2020.

CHAPTER 9
NANOTECHNOLOGY IN SPACE EXPLORATION

"Nanotechnology will let us build computers that are incredibly powerful. We'll have more power in the volume of a sugar cube than exists in the entire world today."

—Ralph Merkle

9.1 INTRODUCTION

Nanotechnology is a rapidly developing field that has the potential to revolutionize many aspects of our lives, including space exploration and colonization. It is a young and dynamic discipline that manipulates matter at the atomic and molecular level to alter its qualities and improve its performance. Although it is still in its early stages of development, it has the potential to revolutionize space exploration and colonization. The main objective of nanotechnology research is to reduce weight and improve the performance and durability of materials, electronics, and power systems. Nanotechnology has the potential to significantly enhance mission capability by enabling, among others, the development of materials that are significantly lighter and have mechanical properties and durability superior to those available today [1].

Space development, the practical application of the capabilities of spacecraft and of the data collected from space, has evolved in parallel with space exploration. Many space applications have both civilian and military uses, and thus similar systems have been developed by both sectors.

Nanotechnology is revolutionizing space travel and reshaping our understanding of the cosmos. The realm of space exploration has always been about pushing boundaries, seeking new frontiers, and advancing technology to achieve the seemingly impossible. Harnessing the power of nanotechnology has the potential to revolutionize space exploration and colonization in ways that were once the stuff of science fiction. Nanotechnology has the potential to greatly enhance the capabilities of future space missions and increase their efficiency and effectiveness [2,3].

This chapter examines the implication of nanotechnology in space and its effects on geopolitics. It begins with describing what nanotechnology is all about. It covers space nanotechnology. It presents some applications of nanotechnology in space exploration. It discusses space nanotechnology around the world. It highlights some benefits and challenges of nanotechnology in space exploration. The last section concludes with comments.

9.2 WHAT IS NANOTECHNOLOGY?

Technologies impact every aspect of our modern society. There are many ways in which our society and technology are interlinked. Nanotechnology has the potential to provide huge benefits, just like any useful technology.

The term “nano” means something small, tiny, and atomic in nature. The application of the term in science led to a field called nanotechnology. Nanotechnology refers to the characterization, fabrication and manipulation of structures, devices or materials that have one or more dimensions that are smaller than 100 nanometers. It may be regarded as an area of science and engineering where phenomena that take place at the nano-scale (10-9m) are utilized in the design, production, and application of materials and systems. It is an emerging area of that integrates chemistry, biology, and materials science to create new properties that can be exploited to gain new market opportunities [4].

Nanotechnology deals with the characterization, fabrication, and manipulation of biological and nonbiological structures smaller than 100 nm. Dimensions between approximately 1 and 100 nanometers are known as the nanoscale. To put this in perspective, a single nanometer is about 100,000 times smaller than the width of a human hair. As indicated in Figure 9.1 [5], the nanoscale is so small that we cannot see it with a light microscope.

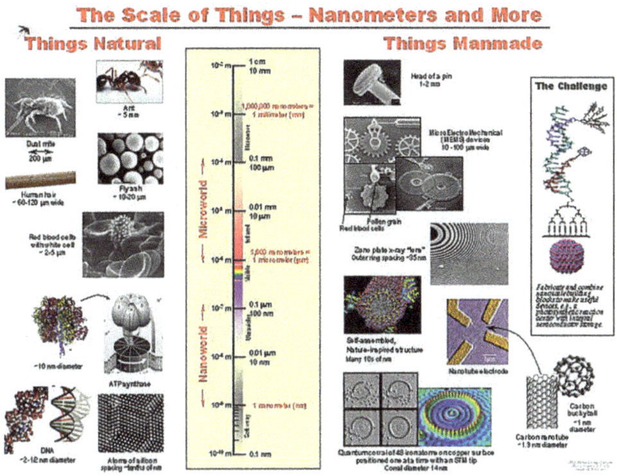

Figure 9.1 Indicating the relative scale of nanosized objects [5].

It is the scale of atoms and molecules. Nanotechnology involves the creation and application of materials and devices at the level of molecules and atoms. It may be regarded as the science that is conducted, researched, investigated, and experimented at the nanoscale. Nanotechnology is a multi-disciplinary field that includes biology, chemistry, physics, material science, and engineering. It is the science of small things—at the atomic level or nanoscale level. The past three decades has witnessed an increased interest and funding in nanotechnology. This has led to rapid developments in all areas of science and engineering [6].

Richard Feymann, the Nobel Prize-winning physicist, introduced the world to nanotechnology in 1959 and is regarded as the father of nanotechnology. Nanotechnology involves the manipulation of atoms and molecules at the nanoscale so that materials have new unique properties. Nanomaterials are expected to have at least one dimension (length, width, height) at the nanoscale of 1 – 100 nm. One nanometer is a billionth of a meter, too small to be seen with a conventional lab microscope. Nanomaterials include nanofilters, nanosensors, nano photocatalysts, and nanoparticles. Nanomaterials are known as nanoparticles when they have nanoscale length, width, and height. Figure 9.2 portrays a technique for the preparation of nanoparticles [7].

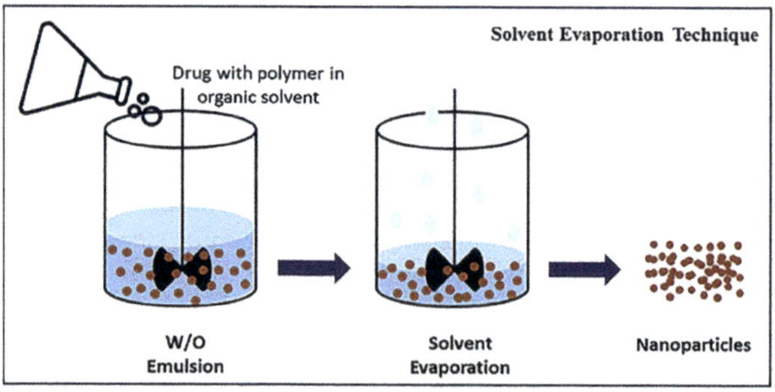

Figure 9.2 A technique for the preparation of nanoparticles [7].

Today, nanotechnology is part of our daily lives. Nanotechnology will leave virtually no aspect of life untouched. Its usages include everything from safer food processing to more efficient drug-delivery systems to powerful computer chips. Three steps to achieving nanotechnology-produced goods are [8]:

1. Scientists must be able to manipulate individual atoms.
2. Next step is to develop nanoscopic machines, called assemblers, that can be programmed to manipulate atoms and molecules at will.
3. In order to create enough assemblers to build consumer goods, some nanomachines called replicators, will be programmed to build more assemblers.

Nanotechnology is trending among scientists and engineers. Here are some underlying trends one should look for [9]:

1. *Stronger Materials:* The next generation of graphene and carbon devices will lead to even lighter but stronger structures.
2. *Scalability of Production:* One big challenge is how to produce nanomaterials that make them affordable. Limited scalability often hinders application.
3. *More Commercialization:* In addition to transforming the automotive, aerospace, and sporting goods fields, nanotechnology is facilitating so many diverse improvements: thinner, affordable, and more durable.

4. *Sustainability:* One main goal of the National Nanotechnology Initiative, a US government program coordinating communication and collaboration for nanotechnology activities, is to find nanotechnology solutions to sustainability.

5. *Nanomedicine:* There will be a mindboggling impact of nanotechnology on medicine, where advances are being made in both diagnostics and treatment areas.

Applications of nanotechnology are found in a wide range of industries, including engineering, medicine, microelectronics, manufacturing, biology, chemistry, energy, and agriculture, and life sciences. Figure 9.3 shows some applications of nanotechnology [10]. Although nanotechnology has been successfully applied in various industries, its use in the oil and gas sector is still limited.

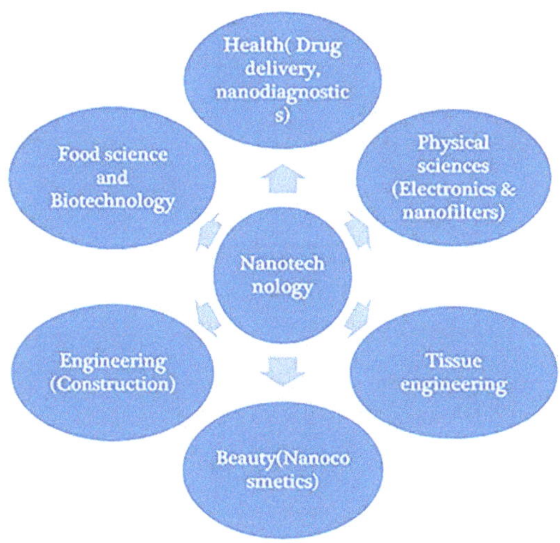

Figure 9.3 Some applications of nanotechnology [10].

9.3 SPACE NANOTECHNOLOGY

Nanotechnology deals with manipulating matter at the nanoscale, typically in the range of 1 to 100 nanometers. It is revolutionizing space exploration by driving breakthroughs in space research. It may hold the key to making spaceflight more practical. The use of nanomaterials, such as carbon nanotubes, is transforming spacecraft design, making them more robust yet weightless. Carbon nanotubes have exceptional mechanical properties that make them ideal for spacecraft components. As shown in Figure 9.4, carbon nanotubes are cylinder-shaped molecules that consist of rolled-up sheets of single-layer carbon atoms [11].

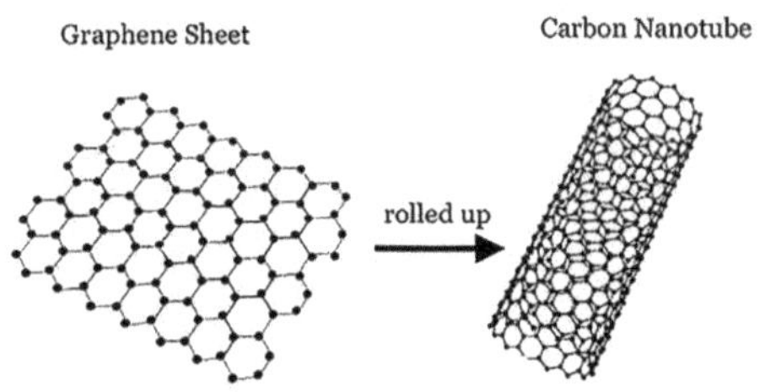

Figure 9.4 Carbon nanotubes are made from graphene sheet [11].

Carbon nanotubes have been used to develop lightweight and strong materials for spacecraft, such as solar sails, which use the pressure of sunlight to propel a spacecraft through space. Since the invention of carbon nanotubes in 1991, there has been an uproar in the scientific community around the various applications of this material.

The exploration of space is no easy task due to its sheer vastness. Space visionaries in the early 20th century recognized that putting satellites into orbit could furnish direct and tangible benefits to people on earth. Modern space age began with the launching of the Sputnik satellite in October of 1957. The Mercury Seven, shown in Figure 9.5, was the first group of American astronauts [12].

Figure 9.5 The Mercury Seven of American astronauts [12].

The goal of Project Mercury was to see if it was possible to land a man on the moon. Neil Armstrong became the first person to leave a footprint on the surface of the moon on July 20, 1969. It has been over 50 years since the first human walked on the Moon. The Voyager is a space probe that was designed to explore the furthest reaches of the Solar System. NASA successfully landed two rovers (robots) on the surface of Mars in January of 2004. International Space Station, shown in Figure 9.6, is a joint project between several countries [12]. It is currently the largest space station ever constructed.

Figure 9.6 International Space Station [12].

9.4 APPLICATIONS OF SPACE NANOTECHNOLOGY

Nanotechnology, the manipulation of materials and devices at the atomic and molecular scale, has the potential to revolutionize space science and exploration. From improving spacecraft materials to enabling resource utilization on distant worlds, the applications are vast and exciting. Nanotechnology is used in many ways in space exploration, including [13,14]:

- *Sensors:* Nanoscale sensors are an exciting application of nanotechnology in space exploration. Nanoscale sensors play a critical role in capturing data and monitoring the space environment. Nanotechnology can be used to create miniature sensors that can collect data and take images in space. Nanosensors can be used to monitor the levels of trace chemicals in spacecraft to monitor the performance of life support systems. Nanosensors can also be used to monitor spacecraft systems and the health of astronauts. By harnessing nanomaterials and nanoscale sensors, we are paving the way for unprecedented achievements in space research. By leveraging nanoscale sensors, scientists and engineers are able to gather data and monitor the space environment with unprecedented precision. Energy fuels sensors, devices, and communication systems, enable astronauts to interact efficiently with their surroundings.

- *Spacecraft:* Nanomaterials can make spacecraft lighter and stronger, which can improve propulsion efficiency and reduce launch costs. Nanomaterials can also be used to shield spacecraft from radiation. Materials made from carbon nanotubes can be employed to reduce the weight of spaceships. NASA is developing a nanomaterial-based coating for spacecraft that can protect them from the harsh space environment. At the moment, 95% of the weight of a spacecraft at launch is fuel, leaving only 5% for the craft itself, payload and astronauts. The development of nanomaterials could lead to lighter and more efficient fuel storage, and the use of nanoscale structures in propulsion systems could increase their performance.
- *Space suits:* Nanomaterials can be used to make spacesuits lighter and more flexible, which can help astronauts explore and maneuver more easily. Nanomaterials can also be used to create self-repairing spacesuits that can seal punctures and protect astronauts from toxic atmospheres. New materials combined with nanosensors and nanorobots could improve the performance of spaceships, spacesuits, and the equipment used to explore planets and moons. There could be layers of bio-nano robots in spacesuits. The outer layer of bio-nano robots would respond to damages to the spacesuit. An inner layer of bio-nano robots could respond if the astronaut was in trouble, for example by providing drugs in a medical emergency. Nano-coatings applied to space suits ensure the safety and comfort of astronauts during their missions.
- *Propulsion:* Propulsion is one of the most significant challenges to developing fast and convenient space travel. Nanoparticles can be used in rocket propellants to improve thrust and fuel efficiency. Nano-engineered propulsion systems, such as ion thrusters, offer significant advantages in space exploration. The integration of nanotechnology will be critical in developing advanced propulsion systems that can revolutionize space travel. Nanotechnology can be

used to create nano-engineered propulsion systems, such as ion thrusters. Nanoparticles can also be integrated into rocket propellants to enhance thrust and fuel efficiency. This breakthrough in propulsion technology will open up new frontiers for human exploration, enabling us to reach further into the depths of our solar system.

- *Space Colonization:* Nanotechnology plays a pivotal role in resource utilization on celestial bodies like the Moon or Mars. In space colonization efforts, life support systems are crucial for maintaining the health and well-being of astronauts. Nanotechnology can contribute to the development of compact, efficient life support systems that are essential for the sustainability of human colonies in space self-replicating nanobots hold promise for space colonization.

- *Energy:* Energy harvesting is another area where nanotechnology plays a vital role in space exploration. Nanotechnology is employed in advanced solar cells, like quantum dot solar cells, which can efficiently convert sunlight into electricity. Nanotechnology is transforming energy harvesting in space exploration. Nanogenerators, powered by piezoelectric nanomaterials, capture energy from vibrations and temperature variations in space. This energy fuels sensors, devices, and communication systems, enabling astronauts to interact efficiently with their surroundings. Nanogenerators powers sensors, communication systems, and other devices, reducing the reliance on traditional energy sources and extending the capabilities of astronauts during their missions. By leveraging nanotechnology in space suit technology and energy harvesting, we are taking significant steps towards enabling longer-duration space missions and ensuring the well-being of astronauts.

- *Nanomedicine:* This offers groundbreaking possibilities for healthcare in space colonization. Nanoparticles can be designed for targeted drug delivery, ensuring that medications reach specific cells or tissues precisely when needed. Advancements in healthcare can enhance the overall quality of life for space colonists and provide essential medical care during extended missions. Researchers at the University of California, Berkeley are developing a nanomaterial-based drug delivery system that can be used to treat astronauts for medical conditions in space. Figure 9.7 shows biological and health features of spaceflight [15].

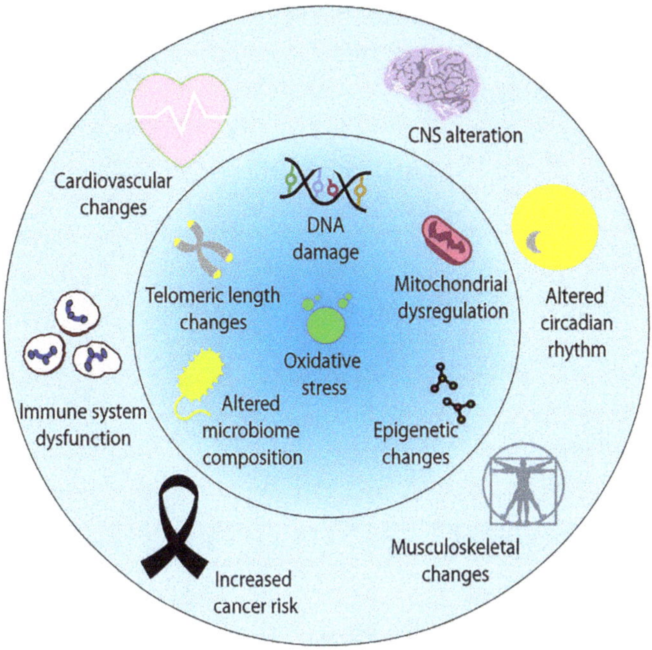

Figure 9.7 Biological and health features of spaceflight [15].

- *Military:* A nation's global power is influenced by several different elements including military prowess, economic strength, diplomatic relations, and soft power. Advancement in technology can also enhance a country's overall development and contribute to a country's hegemony. Some nations and organizations were quick to recognize the great usefulness of space-based systems in military operations. Apart from land, air, and sea, space has become the fourth military dimension. The United States, Russia, the United Kingdom, France, China, the North Atlantic Treaty Organization (NATO), and other European countries have deployed increasingly sophisticated space systems. These nations are working to develop military-grade materials and systems based on nanotechnology. The relationships and rivalry between space-faring nations, nations mold the geopolitical environment of space operations.

- *Materials:* Nanotechnology can create lighter, stronger, and more durable materials for spacecraft components. Development of advanced materials using nanotechnology can improve performance in the following areas: electrical energy generation and storage, propulsion, sensors, instrumentation, signal and power transmission, thermal protection, and active structures sensing, healing, and shape control. NASA's internal development program should focus on the testing and use of nano-enhanced materials and structures and understanding their performance in the extreme environment found in space. NASA is developing carbon nanotube (CNT) reinforced composite overwrap pressure vessels (COPVs). Future NASA missions depend highly on advancements such as lighter and stronger materials, increased reliability, and reduced manufacturing and operating costs

9.5 SPACE NANOTECHNOLOGY AROUND THE WORLD

Historically, government-sponsored spaceflights were mission-based and geared towards national interests, e.g., technological leadership, improving national security, creating high-quality jobs, or advancements in research. Only a few nations, known as the space-faring nations, are predominantly in possession of space capabilities. Such nations, including the US, Russia and China, already have either built or are developing the technology to construct ground-based directed energy weapons. A nation that is behind in nanotechnology will be marginalized in the changing space environment, which could result in geopolitical instability and conflicts. We now consider how some nations are incorporating nanotechnology into the space programs [16-18]:

- *United States:* North America holds the largest market share for nanotechnology, largely attributed to the influence of large research labs. The National Nanotechnology Initiative (NNI) is a US government research and development initiative.

It focuses on applications that include nanomedicine, nanoelectronics, water treatment, precision agriculture, transportation, and energy generation and storage. It ensures that the United States remains not only the place where nanoscience discoveries are made but also where these discoveries are transported and manufactured into products to benefit society. The overall market for nanotechnology in defense and homeland security is expected to reach $9.9 billion by 2028 and the market for nanosensors is expected to reach $21.4 billion by 2028.

- *United Kingdom:* The London Centre for Nanotechnology is a multidisciplinary research center in physical and biomedical nanotechnology in London, United Kingdom. It brings together institutions that are referents in nanotechnology. The center's work requires it to draw on the combined skills of multiple departments, including medicine, chemistry, physics, electrical and electronic engineering, biochemical engineering, materials and earth sciences, and two leading business centers.

- *Russia:* Russia has a long-standing reputation in the areas of science and technology, with most R&D being carried out in state-owned branch research institutes. Since the collapse of the Soviet Union, Russia has introduced drastic economic changes to adapt to global standards and become more market-based. Russia has many organizations and networks committed to promoting and exploring nanotechnology. Russia is home to numerous universities offering research and educational opportunities in nanotechnology. In August 2014, two Russian cosmonauts launched a nano-satellite from the International Space Station (ISS). This was a joint project between the Peruvian Government and the Russian Federal Space Agency. Russia is focused on promoting and showcasing its nanotechnology expertise to the global community.

- *China:* There is no doubt that China is active in scientific development and nanotechnology. China is a leading country in nanotechnology research. China's success in nanotechnology is raising the standard for other globally competitive nations. Research and development in nanotechnology and nanoscience is the key component of the ambitious "Made in China 2025" initiative aiming at making China a high-tech manufacturing powerhouse. In 2020, researchers in China`s state key laboratory of robotics produced a laser that generates a small gas bubble that can be utilized as a nanorobot, designed to move materials at the nanoscale. Figure 9.8 shows the Xuntian space telescope, which is scheduled for launch in 2025 [19].

Figure 9.8 The Xuntian space telescope, which is scheduled for launch in 2025 [19].

- *India:* The Indian government has been actively promoting the development of the space sector, and private companies are playing an increasingly important role. The nanotechnology market in India is expanding quickly, with numerous applications in the healthcare, consumer items, energy, and military sectors. Building upon the promotional activities carried out as part of the Nano Science and Technology Initiative (NSTI) in the highly promising and competitive area of Nano Science and Technology, which was formed in 2001. Recognizing the success of the Nano Mission, the Union Cabinet accorded approval for the continuation of the Nano Mission in its Phase II during the 12th Plan period with an allocation of Rs. 650 crores. The Department of Science and Technology is the nodal agency for implementing the Nano

Mission. India has been increasing its national number of nanotechnology scientific and technical publications. Indian government took special initiatives to provide world-class education and opportunities for research and development to the students. However, India still has a long way to go. More funds should be allocated by the Indian government for further research. Figure 9.9 shows space exploration in India [20].

Figure 9.9 Space exploration in India [20].

- *Iran:* With the support of talented academicians and knowledge-based companies, the nanotechnology sector has indigenized many technologies to solve the main challenges of the country in various areas, including industry. The national document on promoting the application of nanotechnology has outlined the path to achieving the major goals of the development of nanotechnology in the country. According to the document, 12,199 articles by Iranian researchers related to nanotechnology were indexed in the Web of Science (WoS) in 2021. This share of the nanotechnology articles placed Iran fourth in the world in 2020 and 2021. Iran currently ranks fourth in nanotechnology in the world after the United States, India, and China. The expansion of exports in recent years and the creation of bases in China, India, Indonesia, Syria, Turkey, and Iraq have provided a platform for the entry of Iran's nanotechnology goods, equipment, and services into the world markets.

9.6 BENEFITS

Nanotechnology may hold the key to making space flight more practical. Advancements in nanomaterials make lightweight solar sails and a cable for the space elevator possible. By significantly reducing the amount of rocket fuel required, these advances could lower the cost of reaching orbit and traveling in space. Other benefits include [21]:

- *Space Suits:* Nanotechnology could be used to create space suits that provide astronauts with additional protection and can seal and repair breaches.
- *Space Tourism:* Nanotechnology holds great potential in the field of space tourism. As space tourism continues to gain traction, nanotechnology can play a crucial role in developing lightweight and durable spacecraft, enhancing the safety and comfort of future space travelers.
- *Miniaturization:* Nanotechnology can lead to the creation of miniaturized devices and instruments that can be used for space missions. This will result in smaller and lighter payloads, which will reduce launch costs and increase the efficiency of space missions.
- *Improved Materials:* Nanotechnology can lead to the development of new materials with improved properties such as increased strength, durability, and thermal resistance. These materials can be used to construct lighter and more robust spacecraft.
- *Energy Efficiency:* Nanotechnology can contribute to the development of more efficient solar panels and energy storage devices, enabling spacecraft to generate and store more energy. This will increase the autonomy of spacecraft and reduce the need for frequent recharging or refueling.
- *Radiation Shielding:* This is an area where nanotechnology could make a major contribution to human space flight. Human tissue and electronic component protection from the harmful effects of space radiation is essential for extended deep space exploration voyages. NASA says that the risks

of exposure to space radiation are the most significant factor limiting humans' ability to participate in long-duration space missions. A lot of research therefore focuses on developing countermeasures to protect astronauts from those risks. Another area of required radiation shielding is the protection of onboard electronics. It has been reported that electronic devices become more radiation tolerant when their dimensions are reduced.

- *Nanorobots:* Exploring space with a swarm of nanobots is another popular application of nanotech in space. These robots can be very small and grouped to create autonomous nanotechnology swarms (ANTS). Nanobots can effectively cover more ground and explore at much faster rates.
- *Energy:* Energy generation and energy storage will remain a top technical challenge for all future space-related missions. Batteries and power generation account for a significant amount of weight in any launch vehicle. Efficient methods to generate and recover energy, reduce overall power requirements, and reduce weight will benefit future NASA missions. Nanotechnology can improve performance for energy generation, energy storage, and energy distribution.

9.7 CHALLENGES

Space exploration is not without challenges. Space missions aim to reach Mars and beyond, but many challenges must be overcome before this can be achieved.

Safety concerns, ethical dilemmas, and the unknown risks of deploying nanoscale technologies in space must be carefully addressed. Additionally, international cooperation and regulations will be crucial to ensure responsible and equitable use of these technologies. Other challenges include the following [18]:

- *Ethical Concerns:* Space exploration brings with it an increasingly diverse set of space travelers, raising new and nontrivial ethical, legal, and medical policy and practice concerns which are still relatively underexplored. Commercial spaceflight raises new ethical considerations.

Principles of health ethics encompass avoiding harm, beneficence, achieving a favorable risk-benefit balance, respecting autonomy, fairness, and fidelity.

- *Health Risks:* Health concerns regarding commercial space travel are of great importance considering that the aforementioned potential hazards can severely interfere with many physiological processes. Exposure to galactic cosmic rays and solar particle events is potentially the most significant single health hazard. Recently, NASA increased an individual astronaut's total career effective radiation dose (independent of age at exposure and sex) due to spaceflight radiation exposure to less than 600 mSv. Figure 9.10 depicts the five hazards in space contributing to increased health risks [15].

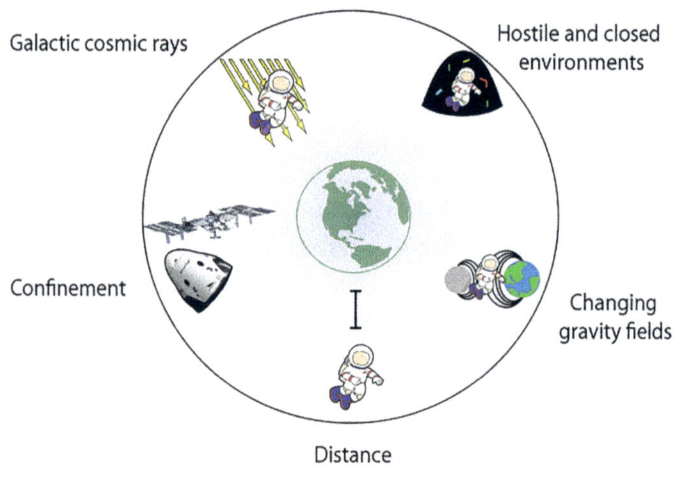

Figure 9.10 Five hazards in space contributing to increased health risks [15].

- *Space Radiation:* NASA believes the risk of exposure to space radiation is the most significant issue restricting humans' capacity for long-duration spaceflight. Nanotechnology could contribute to improving the spacecraft themselves, while also protecting both astronauts and equipment. For space travel, participants must have a good understanding of the space environment, the nature of associated environmental hazards.

- *Space Race:* The space race promoted exploration and use of outer space including the moon, Mars, and other celestial bodies as well as technological development
- *Global Tensions:* Rapid advancements in space technology have been accompanied by an increase in global tensions between world powers, which was a distinguishing feature of the Cold War since it was the first time mankind sought to compete in the space arena. It became a matter of pride for both the US and the erstwhile USSR of their control over space and so technological superiority was necessary to control its national security. After two space-faring nations the US and the USSR, China, Japan, France, the European Space Agency, and Israel joined the club.
- *Regulation:* As new technology emerges, there is a chance of missing it and causing crimes, which creates a need to alter the nature of warfare and requires new negotiation of standards of arms control and compliance with international law. The science of nanotechnology is constantly developing, and new advancements can have impacts in ways hitherto unknown. To ensure security in the future, it is vital to have more understanding and regulation. Some of the regulatory authorities of the United States and the European Union have started assessing the potential risks of nanoparticles.

9.8 CONCLUSION

Nanotechnology is working with things at the nanoscale. The exciting thing about nanotechnology and nanomaterials is that they begin to exhibit weird and unique properties which can be very useful especially for space exploration. By leveraging the properties of nanomaterials we can create lighter and stronger spacecraft. We are currently witnessing the remarkable impact of nanotechnology in pushing the boundaries of exploration. Nanotechnology may hold the key to making space flight more practical.

Looking ahead, the future of nanotechnology in space exploration holds immense promise and is filled with possibilities. It will play an important role in future space missions. More information about nanotechnology and nanomaterials in the space exploration and colonization can be found in the books in [22-24] and the following related journals/magazines:

- *Nanotechnology*
- *Nanoscale.*
- *Nano: The Magazine for Small Science*
- *Micro and Nano Technologies*
- *Nanotechnology News*
- *Nature Nanotechnology*
- *Current Research in Nanotechnology*
- *American Journal of Nanotechnology & Nanomedicine*
- *Nanomedicine: Nanotechnology, Biology and Medicine*
- *Journal of Nanotechnology*
- *Journal of Nanoparticle Research*
- *Journal of Bioelectronics and Nanotechnology*
- *Journal of Nanoscience and Nanotechnology,*
- *Journal of Micro and Nano-Manufacturing*
- *Journal of Nanoengineering and Nanomanufacturing*
- *Nanotechnology and Precision Engineering*

REFERENCE

[1] "Nanotechnology," December 2020

https://www.nasa.gov/directorates/stmd/game-changing-development-program/nanotechnology/

[2] M. N. O. Sadiku, G. K. Suman, and S. M. Musa, "Nanotechnology in space," *Journal of Scientific and Engineering Research*, vol. 8, no. 3, 2021, pp. 166-171.

[3] M. N. O. Sadiku, P. A. Adekunte, and J. O. Sadiku, "Nanotechnology in the space exploration," *International Journal of Trend in Scientific Research and Development*, vol. 8, no. 5, September-October 2024, pp. 965-975.

[4] M. N. O. Sadiku, M. Tembely, and S.M. Musa, "Nanotechnology: An introduction," *International Journal of Software and Hardware Research in Engineering*, vol. 4, no. 5, May. 2016, pp. 40-44.

[5] "Nanotechnology white paper,"

https://www.epa.gov/sites/default/files/201501/documents/nanotechnology_whitepaper.pdf

[6] M. N. O. Sadiku, Y. P. Akhare, A. Ajayi-Majebi, and S. M. Musa, "Nanomaterials: A primer," *International Journal of Advances in Scientific Research and Engineering*, vol. 7, no. 3, March 2020, pp. 1-6.

[7] U. A. Ali et al., "A state-of-the-art review of the application of nanotechnology in the oil and gas industry with a focus on drilling engineering," *Journal of Petroleum Science and Engineering*, vol. 191, August 2020.

[8] K. R. Saravana and R. Vijayalakshmi, "Nanotechnology in dentistry," Indian Journal of Dental Research, November 2005.

[9] N. S. Giges, "Top 5 trends in nanotechnology," March 2013,

https://www.asme.org/topics-resources/content/top-5-trends-in-nanotechnology

[10] D. E. Effiong et al., “Nanotechnology in cosmetics: Basics, current trends and safety concerns—A review,” *Advances in Nanoparticles*, vol. 9, 2020, pp. 1-22.

[11] . Pathania, “Using nanotechnology for space exploration,” February 2020,

https://sumeetpathania2003.medium.com/using-nanotechnology-for-space-exploration-195c1c2fc331

[12] R. Mayda, “Space exploration pictures,”

https://science.howstuffworks.com/space-exploration-pictures.htm

[13] “Nanotech in space exploration: Pushing boundaries beyond earth,” November 2023,

https://www.nanomuscle.com/nanotechnology-in-space-exploration/#:~:text=By%20harnessing%20nanomaterials%20and%20nanoscale,the%20possibilities%20of%20extraterrestrial%20exploration.

[14] R. Uppal, “Nanotechnology in space exploration and colonization: A glimpse into the future,” February 2024,

https://idstch.com/space/nanotechnology-in-space-exploration-and-colonization-a-glimpse-into-the-future/

[15] A. Seylani et al., “Ethical considerations for the age of non-governmental space exploration,” *Nature Communications*, vol. 15, June 2024.

[16] S. Muthuraj, “Nanotechnology for aerospace and associated geopolitics,” August 2023,

https://dras.in/nanotechnology-for-aerospace-and-associated-geopolitics/#:~:text=Introducing%20nanotechnology%2C%20a%20young%20and,paradigm%20shift%20in%20world%-20geopolitics.

[17] “London Centre for Nanotechnology,” *Wikipedia*, the free encyclopedia,

https://en.wikipedia.org/wiki/London_Centre_for_Nanotechnology

[18] "Nanotechnology in Russia: Market report," June 2015,

https://www.azonano.com/article.aspx?ArticleID=3543

[19] A. Jones, "China is embarking on a science and exploration program of staggering scale,"

https://www.astronomy.com/space-exploration/china-deep-space-ambition/

[20] S. Jagriti, "Space technology opportunity in India," January 2024,

https://www.linkedin.com/pulse/space-technology-opportunity-india-globallaunchbase-ypqzc

[21] "Nanotechnology in space," Unknown Source.

[22] M. N. O. Sadiku, S. M. Musa, T. J. Ashaolu, and J. O. Sadiku, *Applications of Nanotechnology*. Gotham Books, 2023.

[23] D. Newberry, Nanotechnology: *Applications to Space Exploration*. Springer, 2024.

[24] M. L. Terranova and E. Tamburri (eds.), *Nanotechnology in Space. Routledge*, 2021.

CHAPTER 10
3D PRINTING IN SPACE

"3D printing represents the democratization of manufacturing. It allows anyone to create anything, anywhere, at any time."

—Neil Harbisson

10.1 INTRODUCTION

Space exploration and inhabiting other celestial bodies are important milestones in technological progress. There are obstacles that prevent humans from simply settling on the Moon and Mars. One promising way to overcome the obstacles is the use of additive manufacturing (AM), also known as 3D printing. Humanity has been using 3D printing in the space industry since the 2000s. In 2011, European aerospace giant Airbus sent the first 3D printed part outside of our atmosphere. The company repeated the feat in 2015, when it produced the first space-qualified 3D printed to replace a satellite component. In 2024, the International Space Station (ISS), the National Aeronautics and Space Administration (NASA), and Made in Space made history by successfully 3D printing the first object in space. These initial successes paved the way for long-term space expeditions.

With the increasing ambitions of deep space exploration, extending to celestial bodies such as the Moon and Mars, there is an unprecedented demand for innovative manufacturing solutions. Many crucial elements for astronaut well-being, such as food, daily necessities, and lunar living accommodations, have the potential to be fabricated through 3D printing. Both NASA and the European Space Agency (ESA) are pushing ahead with plans for 3D printing on the moon and the International Space Station (ISS). Their goal is to build housing, infrastructure, tools, or spare parts that are needed to advance space exploration, possibly using space dust or rock as a raw material. Figure 10.1 shows the world's largest 3D printing facility producing rocket [1]. Apart NASA, ISS, ESA, companies interested in 3D printing in space include Relativity Space, Orbital Matter, Rocketdyne, Boeing, Airbus, Made in Space, and SpaceX.

Figure 10.1 The world's largest 3D printing facility producing rocket [1].

3D printing, also known as additive manufacturing, is a technique for making a three-dimensional object of almost any shape through additive processes in which successive layers of material are laid down under computer control to generate the precise shape of the programmed model. It is an industrial process that has opened up new ways of looking at how parts are designed. It has emerged as a leading in-situ space manufacturing innovation because it offers unprecedented on-demand and rapid manufacturing capabilities in space. 3D printing in space is essential because it can manufacture intricate and customized structures locally, substantially decreasing mission expenses by launching only raw materials and essential components. 3D printing excels at manufacturing components for emergency repairs in space [2].

This chapter examines the use of 3D printing technology in space. It begins with explaining what 3D printing is all about. It presents 3D printing in the space industry. It provides several applications of 3D printing in space. It highlights the benefits and challenges of employing 3D printing in space industry. It concludes with some comments.

10.2 WHAT IS 3D PRINTING?

3D printing (also known as additive manufacturing (AM) or rapid prototyping (RP)) was invented in the early 1980s by Charles Hull, who is regarded as the father of 3D printing. Since then it has been used in manufacturing, automotive, electronics, aviation, aerospace, aeronautics, engineering, architecture, pharmaceutics, consumer products, education, entertainment, medicine, space missions, the military, chemical industry, maritime industry, printing industry, and jewelry industry [3]

A 3D printer works by "printing" objects. Instead of using ink, it uses more substantive materials–plastics, metal, rubber, and the like. It scans an object–or takes an existing scan of an object–and slices it into layers, which can then convert into a physical object. Layer by layer, the 3D printer can replicate images created in CAD programs. In other words, 3D printing instructs a computer to apply layer upon layer of a specific material (such as plastic or metal) until the final product is built. This is distinct from conventional manufacturing methods, which often rely on removal (by cutting, drilling, chopping, grinding, forging, etc.) instead of addition. Models can be multi-colored to highlight important features, such as tumors, cavities, and vascular tracks. 3DP technology can build a 3D object in almost any shape imaginable as defined in a computer-aided design (CAD) file. It is additive technology as distinct from traditional manufacturing techniques, which are subtractive processes in which material is removed by cutting or drilling [4].

3D printing has started breaking through into the mainstream in recent years, with some models becoming affordable enough for home use. Many industries and professions around the world now use 3D printing. It plays a key role in making companies more competitive. The gap between industry and graduating students can be bridged by including the same cutting-edge tools, such as 3D printing, professionals use every day into the curriculum. There are 3D printed homes, prosthetics, surgical devices, drones, hearing aids, and electric engine components. As shown in Figure 10.2, 3D printing involves three steps [5]. A typical 3D printer is shown in Figure 10.3 [6].

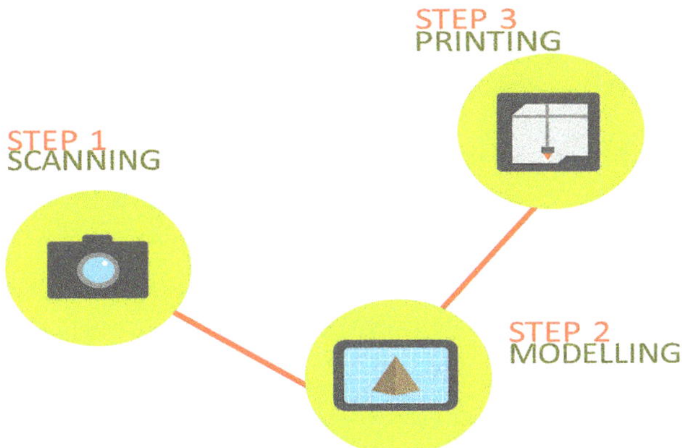

Figure 10.2 3D printing involves three steps [5].

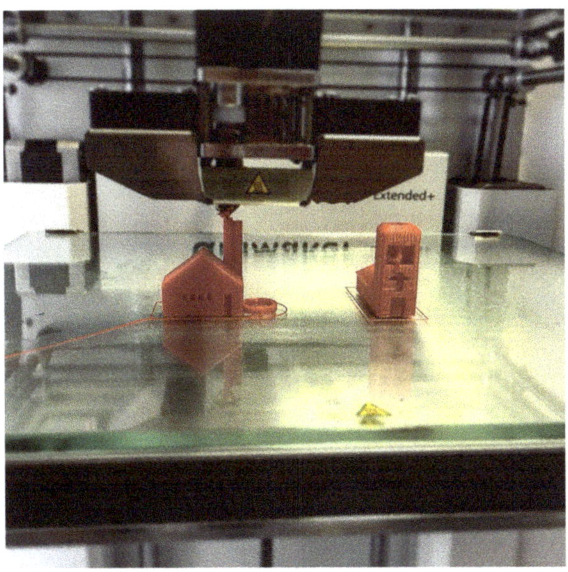

Figure 10.3 A typical 3D printer [6].

10.3 3D PRINTING IN SPACE

The process of 3D printing involves the creation of three-dimensional objects by layering materials according to a digital model. This cutting-edge technology has found its way into diverse fields, ranging from healthcare to construction. It is already disrupting many fashion, engineering, and aerospace industries. It is often much cheaper than traditional industrial manufacturing methods, especially in low-volume applications. Today, 3D printing is increasingly becoming adopted in the space sector. 3D printing in space starts with getting off the ground and escaping Earth's gravity.

To date, 3D printing has revealed promising opportunities across diverse space applications, including the production of space devices and food, advancements in space biomedicine, repairs of electronics and sensors, and the recovery and utilization of space resources. Standard 3D printers use plastics or metals to construct three-dimensional objects. NASA, the US space agency, is a pioneer in 3D printing exploration, providing funding for many new technologies. Its enthusiastic adoption of 3D printing is largely due to cost. 3D printing provides the low production volumes, high degrees of design freedom, and low cost that their missions require. In 2013, NASA announced funding for the development of a 3D printer that could create food for the astronauts. It succeeded in creating 3D print pizza in space, as shown in Figure 10.4 [7].

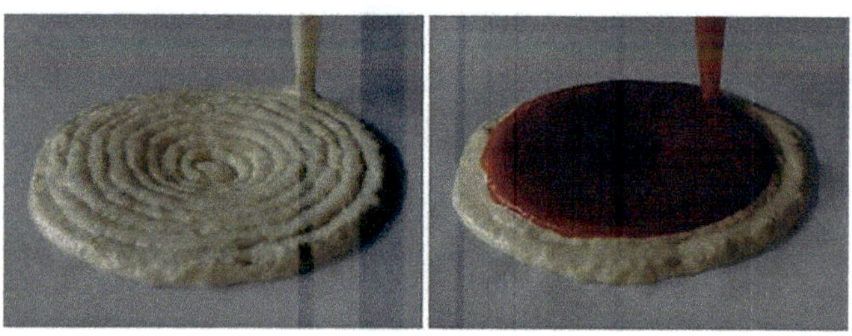

Figure 10.4 3D print pizza in space [7].

3D printing in space is an impressive technological feat that has yielded significant achievements. Sophisticated computer systems can help mission control systems on Earth to remotely operate 3D printers aboard spacecraft or present in space stations. Therefore, without advanced computing systems, the entire process of 3D printing, both on Earth and in space, would not be viable [8].

Regarding materials, the existing in-space 3D printing technology can be broadly categorized into three categories: polymers, metals, and cell-laden substances. Polymer 3D printing in space has evolved significantly, demonstrating technological advancements and the successful on-demand manufacturing of functional components. Metal 3D printing in space originated from the manufacturing of tools and will bring new on-orbit manufacturing capabilities.

10.4 APPLICATIONS OF 3D PRINTING IN SPACE

The 3D printing technology minimizes waste and continues to advance rapidly in terms of size, cost, complexity, and types of printable materials. It has many uses from the day-to-day, from printing whole houses to producing spacecraft parts. The following are typical applications of 3D printing in space exploration.

1. *Astronauts:* Astronauts benefit from 3D printing with devices aboard spacecraft and stations that help to sustain long missions in space between resupplying. Current astronauts contend with in weightlessness, tackled in part through exercise and medical countermeasures. This might be made more serious still by spending years rather than months in orbit. Astronauts have already used 3D printers to replace or repair plastic parts. On-demand manufacturing allows astronauts to quickly produce tools and components tailored to their specific needs. Figure 10.5 shows some astronauts [9].

Figure 10.5 Some astronauts [9].

2. *Bioprinting in Space:* Prolonged exposure to the space environment, marked by microgravity and increased radiation, leads to detrimental effects such as muscle and bone mass loss,

reduced cardiovascular activity, altered motor functions, and impaired wound healing. To comprehend these impacts and develop patient-specific tissue constructs for astronauts, 3D bioprinting has emerged as a powerful tool. A 3D bioprinting capability would allow a more flexible, versatile response to emergencies.

3. *International Space Station (ISS):* This has continuously been home to astronauts for more than nineteen years. Astronauts conduct scientific research using dozens of special facilities aboard the space station, which also provides them with a place to eat, sleep, relax, and exercise. To make all of this possible requires sending more than 7,000 pounds of spare parts to the station annually. The project sent the first 3D printer to the space station in 2014. Developed by Made in Space, this printer used a fused filament fabrication (FFF) process, feeding a continuous thread of plastic through a heated extruder and onto a tray layer by layer to create a 3D object. This particular 3D printing process is not affected in a significant way by microgravity [10].

4. *A Zero-Gravity 3D Printer:* Using the 3D printing technology, it is now possible to manufacture spare parts in space. Astronauts have found that additive manufacturing is a viable technology to use even in microgravity. NASA and Made in Space collaborated to produce to a 3D printer that can operate in zero-gravity environments. Figure 10.6 shows the Zero G 3D printer by Made in Space in collaboration with NASA [7].

Figure 10.6 The Zero-G 3D printer [7].

5. *Satellites:* 3D printing is also increasingly being used in space is for satellites. Satellites are increasingly being made using additive manufacturing in space. Currently, there are projects from a number of companies including Boeing and Airbus which have used additive manufacturing to create increasingly complex, lighter parts for their satellites.

6. *Aerospace:* 3D printing technology is increasingly being used in aerospace and particularly space exploration for various applications. This is because it allows for lighter, cheaper, and more efficient parts to be made due to the unique geometries that are possible with AM as well as the materials used. One such application involves using additive manufacturing for thrusters. Additive manufacturing is also being used in the aerospace industry for the creation of 3D printed rockets. The aerospace industry is facing is the creation of 3D printed food.

7. *Rocket Engines:* Space companies are turning to 3D printing. More and more often, engineers are using the SLS process to 3D print rocket engine parts with a copper alloy powder, which can withstand high temperatures. They have designed a 3D printed rocket engine that is 15% more efficient than those traditionally produced. Figure 10.7 shows 3D printed

rocket engine by SpaceX [7].

Figure 10.7 3D printed rocket engine by SpaceX [7].

8. *In-space Manufacturing:* For long space missions, it might be convenient to have a way to manufacture spacecraft parts in case unexpected damage occurs. Currently, the biggest use of 3D printing in the space industry is earth-based manufacturing of spacecraft parts. For example, SpaceX uses 3D printing to produce parts for its Falcon 9, Dragon, and Starship spacecraft. Some companies seek to manufacture entire rockets and spacecraft using 3D printing. Efforts are underway to advance 3D printing to support in-space manufacturing, a capability that will become increasingly important for the future of space travel. It will reduce the need for costly, resource-intensive resupply missions to space stations and other off-world bases. It will also lead to the establishment of space-based manufacturing platforms (in-orbit, on the Moon, and on planets like Mars). 3D printing in space is an experimental technology that holds vast potential for revolutionizing space exploration by enabling astronauts to manufacture spare parts, tools, key components, and building materials on demand [11]. Companies are interested in establishing in-space manufacturing facilities to take advantage of the zero-gravity and vacuum environment.

9. *Food in Space:* 3D printing could be used to create food for long-duration missions that maintains flavor and nutrients without degrading. The development of food that can be 3D printed is another difficulty that the aerospace sector is confronting. Some companies use 3D technology to make the so-called space food. They are able to generate cell-based meat that has the same texture, flavor, and structure as a traditional steak. An integrated approach to human nutrition in space is being developed by a group of professionals in the fields of technology, food, and space [12]. Space food is a type of food product created and processed for consumption by astronauts during missions to outer space. Such food has specific requirements to provide a balanced diet and adequate nutrition for individuals working in space. An example of space food is displayed in Figure 10.8 [13].

Figure 10.8 An example of space food [13].

10.5 BENEFITS

The advantages of 3D printing in space are many. Advancements in 3D printing technology offer solutions to improve food options, reduce waste, and enable on-demand preparation in space. 3D printing could revolutionize space exploration by helping astronauts create objects in space and on-demand. Even though the raw material still needs to be launched, printing the part is still more efficient than transporting it whole up to its final destination. The flexibility and rapid availability of 3D printing will greatly improve astronauts' autonomy. Other benefits of 3D printing in space include the following [14-17]:

1. *Rapid Prototyping:* 3D printing facilitates rapid prototyping, enabling engineers to produce and test designs quickly. This iterative process allows for more efficient development cycles, reducing the time it takes to refine and perfect components for space missions. 3D printing can manufacture parts within hours, which speeds up the prototyping process. A functional prototype created by 3D printing would save a lot of time. 3D printing can be a cost-effective solution for inexpensive components in relatively small quantities.

2. *Waste:* The additive nature of 3D printing means little waste is generated, eliminating the problem of waste disposal in space. The printer used for space by NASA was developed to test technology for recycling waste plastic materials into high quality 3D-printer filament. Recently, the Made in Space Recycler launched to the space station for an investigation into which materials are most effective for recycling into 3D printing filament and which ones can hold up over multiple uses without degrading.

3. *Savings:* So having a 3D printer on the space station could make crewmembers' lives easier and result in significant savings. NASA has been testing rocket-engine parts made via 3D printing, which is significantly cheaper and more efficient than traditional manufacturing techniques for such parts. Manufacturing in space can reduce the cost and time of transporting materials from earth, allowing for more sustainable and efficient space exploration.

4. *Customization:* 3D printing has the potential to revolutionize manufacturing in space, making it more sustainable, cost efficient, and flexible. It can create customized tools and components for specific missions, which can adapt to changing needs. The technology is capable of building custom parts and objects on demand. Manufacturing in space

can enable the production of custom made parts that are specifically designed for space environments, reducing the risk of equipment failure. The option of printing replacement tissue in a customized shape to directly replace a cancer-affected part of the body is opening up treatment options not only for space exploration but also back on Earth.

5. *Reduced Complexity:* 3D printing small parts of large space structures on the ground and assembling them in space can reduce the complexity of space manufacturing.

6. *Reducing Launch Weight:* A lot of thought goes into getting something ready to head into orbit. It takes a significant amount of fuel to break free of Earth's gravity and lift these massive constructs into orbit. The rocket must generate enough force to exceed the pull of gravity. The heavier the missile, the more difficult that becomes. 3D printing can help reduce launch weight in several ways. 3D-printed rockets are lighter by nature, requiring less fuel and force to escape Earth's gravity. This, in turn, helps reduce the cost of each launch and the amount of fuel burned.

7. *Restocking Supplies:* Packing for a trip is always nerve-wracking. Imagine how much more challenging that becomes when you cannot just run out to the corner store because you are in orbit. Even the best-laid plans never survive contact with the vacuum of space. Astronauts might need something that no one thought to pack. In 2014, International Space Station (ISS) astronauts manufactured the first 3D printed object in space. NASA engineers developed it on the ground to eliminate the need for an extra rocket launch to bring them a single tool. Astronauts can also use 3D printers to build their space suits and clothes. This eliminates the need for re-supplying the spacecraft with new equipment

8. *Recycling in Space:* One problem with sending humanity into outer space is that we tend to generate trash wherever we go. The International Space Station is no exception. 3D printing gives astronauts a way to turn their plastic waste into something useful by recycling it into 3D printing filament.

This means that there is less waste that needs to be discarded or returned to Earth, keeping things cleaner for everyone.

9. *On-Demand Manufacturing:* Additive manufacturing enables the production of parts on-demand, eliminating the need for extensive inventories of spare components. This is particularly important for long-duration missions where resupplying or accessing Earth is not feasible.
10. *Environmentally Friendly:* As this technology reduces the amount of material wastage used this process is inherently environmentally friendly. However, the environmental benefits are extended when you consider factors such as improved fuel efficiency from using lightweight 3D printed parts.

10.6 CHALLENGES

While the process of 3D printing has been mastered on Earth, printing metal in space presents its own set of technical challenges. Challenges of 3D printing in extreme space environments include microgravity, high levels of vacuum, significant temperature differentials, and the impact of cosmic and solar radiation. 3D printing addresses challenges such as mass reduction, intricate component fabrication, and resource constraints. One of the main challenges that astronauts face is the lack of specific replacement parts on-board the space station. Extravehicular 3D printing and planetary 3D printing pose notably more significant challenges than intravehicular 3D printing in space. Other challenges of 3D printing in space include the following [14-18]:

1. *High Cost:* One of the major challenges is the cost of transporting materials and equipment to space. The cost of launching a single kilogram of material into space is currently around $10,000, making it cost-prohibitive to transport large amounts of materials and equipment. The hardest and most expensive thing about space is getting there. So the Pentagon is looking at ways to reduce the bill, including putting 3D printers into orbit to repair smaller satellites.
2. *Microgravity:* One of the challenges space exploration faces

is the absence of gravity in space. This poses formidable challenges to space-based 3D printing, affecting both the printing and post-production phases. The microgravity environment can affect the quality and accuracy of 3D printed objects. The absence of gravity leads to molten materials forming spherical droplets during printing, causing disruptions and compromising print quality. Altered heat flow due to microgravity introduces thermal issues, affecting part quality in fluctuating space temperatures. Getting printers to function in a near-zero gravity environment and with the absence of pressure is not a trivial task. Figure 10.9 shows a 3D printer passing microgravity flight tests [20].

Figure 10.9 A 3D printer passing microgravity flight tests [20].

3. *Vacuum of Space:* The vacuum of space, characterized by the absence of an atmosphere, fundamentally alters the heat transfer dynamics essential for the 3D printing process. The lack of convection as a heat dissipation mechanism in a vacuum requires innovative thermal management strategies. The vacuum environment introduces challenges in layer deposition.

4. *Suitable Materials:* One of the major disadvantages is the limited availability of suitable materials in space for 3D printing. There are limits on the amount of materials that can be transported into space. This has led to the development of innovative printing techniques and materials capable

of utilizing indigenous resources. In space, where there is microgravity or zero gravity, the behavior of materials is different

5. *Safety:* This involves protecting the International Space Station (ISS) from the aggressive printing environment caused by the laser and the heat it generates. The printer sits in a sealed metal box, which acts like a safe. Safety and contamination are key drivers for not only for the ISS, but for future use on the Moon.
6. *No Standards:* One challenge is that there really are not firm standards for how to test 3D printed materials, but various organizations, including NASA, are working on test procedures.
7. *On-orbit Inspection:* Another challenge is on-orbit inspection of parts made in space. To use a part in space, you need to know that it meets the requirements for its intended use. Testing these manufacturing systems on the space station paves the way to allow those missions to be more independent of Earth.
8. *Large Volumes:* 3D printing is a static cost unlike more conventional techniques like injection moulding, where large volumes may be more cost effective to produce. While the initial investment for 3D printing may be lower than other manufacturing methods, once scaled up to produce large volumes for mass production, the cost per unit does not reduce as it would with injection moulding.
9. *Reduction in Manufacturing Jobs:* Another of the disadvantages of 3D technology is the potential reduction in human labor, since most of the production is automated and done by printers. However, many third world countries rely on low skill jobs to keep their economies running, and this technology could put these manufacturing jobs at risk by cutting out the need for production abroad.
10. *Design Inaccuracies:* Another potential problem with 3D printing is directly related to the type of machine or process

used, with some printers having lower tolerances, meaning that final parts may differ from the original design. This can be fixed in post processing, but it must be considered that this will further increase the time and cost of production.

Overcoming these technical hurdles is crucial to ensuring the reliability and precision of 3D printing processes in space. Given the rate of progress so far, there is confident that these challenges will be overcome.

10.7 CONCLUSION

NASA, the European Space Agency (ESA), and the space agencies of Russia, China, and India are all actively exploring, and already using 3D printing to help meet their exploration, scientific, security, and commercial goals. NASA is working to develop and incorporate 3D printing technology on several different fronts. For example, the agency awarded a $125,000 grant to a researcher seeking to build a prototype 3D printer for food. ESA plans to launch its own 3D printer to the International Space Station in the first half of 2010. ESA is recently investigating the feasibility of using 3D-printing technology to build a moon base using lunar materials. 3D printing offers a potential means of facilitating lunar settlement with reduced logistics from Earth [21]. SpaceX is another company using 3D printing as an integral part of manufacturing.

3D printing has moved from a hobbyist's best friend to one of the most useful tools in nearly every industry. In space exploration, 3D printing has simply become part of life. The idea of manufacturing goods in space has gained attention. The marriage of 3D printing and space exploration represents a frontier of innovation reshaping how we conceive and execute missions beyond our home planet. 3D printing is revolutionizing launch vehicle design by reducing component counts, integrating complex geometry into lightweight structures, and speeding the whole design and manufacturing cycle. The predictions for the 2020s are high: Space tourism will become a reality, the first woman will reach the Moon, a permanent lunar base will be built, exploration of our solar system will boom, the first human mission will get to Mars, and commercial manufacturing in space will begin. Once in space, 3D printing has several key roles to play: The "zero-g" environment of space orbit opens doors to print materials not possible under the influence of earthly gravity [1]. More information on 3D printing in the space industry is available in the books in [22-26].

REFERENCE

[1] "3D printing in space: 10+ projects to watch in 2021," January 2021,

https://all3dp.com/2/3d-printing-in-space-projects/#google_vignette

[2] M. N. O. Sadiku, U. C. Chukwu, and J. O. Sadiku, "3D printing in space," *Innovative Mult-disciplinary Journal of Applied Technology*, vol. 2, no. 5, 2024, pp. 238-248.

[3] F. R. Ishengoma and T. A. B. Mtaho, "3D printing: Developing countries perspectives computer engineering and applications," *International Journal of Computer Applications*, vol. 104, no. 11, October 2014, pp. 30-34.

[4] M. N. O. Sadiku, S. M. Musa, and O. S. Musa, "3D printing in the chemical industry," *Invention Journal of Research Technology in Engineering and Management*, vol. 2, no. 2, February 2018, pp. 24-26.

[5] D. Pitukcharoen, "3D printing booklet for beginners,"

https://www.metmuseum.org/-/media/files/blogs/digital-media/3d printingbookletforbeginners.pdf

[6] "3D printing in architecture," May 2020,

https://portella.com/blog/3d-printing-in-architecture/

[7] "3D printing for space," August 2021,

https://www.aniwaa.com/guide/3d-printers/3d-printing-for-space/#:~:text=3D%20printing%20could%20revolutionize%20space,things%20like%20food%20or%20buildings!

[8] "3D printing in space: Pioneering humanity's journey beyond earth," April 2024,

https://aapkiawaaz.pk/3d-printing-in-space-pioneering-humanitys-journey-beyond-earth/#:~:text=The%20advantages%20of%203D%20printing,tailored%20to%20 their%20specific%20needs.

[9] D. Jamie, "What are the applications for 3D printing in space?" January 2022,

https://www.3dnatives.com/en/top-10-3d-printing-space/

[10] "Solving the challenges of long duration space flight with 3d printing," December 2019,

https://www.nasa.gov/missions/station/solving-the-challenges-of-long-duration-space-flight-with-3d-printing/

[11] C. Hall, "3D Printing & the Future of Space Exploration," February 2024,

https://www.cutter.com/article/3d-printing-future-space-exploration

[12] "Role of 3D printing in space research," January 2023

https://3dincredible.com/role-of-3d-printing-in-space-research/

[13] "Space food tray,"

https://www.nasa.gov/johnson/exhibits/space-food-tray/

[14] M. Mao et al., "3D printing in space: From mechanical structures to living,′ *International Journal of Extreme Manufacturing*, vol.6, February 2024.

[15] "The world's first metal 3D printer for space is on its way to the ISS," January 2024,

https://www.airbus.com/en/newsroom/stories/2024-01-the-worlds-first-metal-3d-printer-for-space-is-on-its-way-to-the-iss#:~:text=The%20first%20metal%203D%20printer,to%20the%20Moon%20or%20Mars.

[16] "3D printing in space: Advancements and challenges for additive manufacturing,"

https://www.linkedin.com/pulse/3d-printing-space-advancements-challenges-additive

[17] What Are The Benefits of 3D Printing in Space?" April 2023,

https://astroblog.cosmobc.com/benefits-3d-printing-space/

[18] E. Calandrelli, "NASA is sending a 3D printer to space that you can use," March 2016,

https://techcrunch.com/2016/03/19/nasa-is-sending-a-3d-printer-to-space-that-you-can-use/

[19] What are the advantages and disadvantages of 3d printing?

https://www.twi-global.com/technical-knowledge/faqs/what-is-3d-printing/pros-and-cons

[20] "3D printing in space: A new dimension (photo gallery)," August 2013,

https://www.space.com/22359-3d-printing-space-manufacturing-photos.html

[21] M. Wall, "How 3D printing could aid space exploration," November 2014

https://www.space.com/27860-3d-printing-space-exploration.html

[22] M. N. O. Sadiku, *3D Printing and Its Applications*. Moldova, Europe: Lambert Academic Publishing, 2024.

[23] Y. Li, X. Li, and D. Shen, *3D Printing in Space: An Illustrated Introduction*. Singapore: Springer Nature, 2024.

[24] R. Elhajjar and T. Gill, *Studies Into Additive Manufacturing for In-Space Manufacturing*. SAE International, 2010.

[25] J. Yang, *3D Modeling with Tinkercad for 3D Printing - to Space: Just Create*. Toysinbox, 2009.

[26] National Research Council (U.S.). *Committee on Space-Based Additive Manufacturing*, 3D Printing in Space. National Academies Press, 2014.

CHAPTER 11
CYBERSECURITY IN SPACE EXPLORATION

"Security isn't something you buy, it's something you do, and it takes talented people to do it right."

—Anonymous

11.1 INTRODUCTION

Mankind has always been reaching for the stars. Humans have never faltered in going beyond what was thought possible. Space exploration is one of the latest innovative and rapidly growing sectors of human intelligence. It has become increasingly vital to our technological, scientific, and national security interests. It encompasses everything from satellite communications to interplanetary exploration. From telecommunications satellites and GPS to internet connectivity and weather tracking, technology in space is crucial for global communication and innovation [1].

Historically, spacecraft had been considered relatively safe from cyber threats. Space system vulnerabilities were often overlooked in evaluation of critical infrastructure.

Today, satellites and space technologies are now vulnerable to cyber attacks and espionage, such as data manipulation and signal interference. Cybersecurity in space exploration is a critical concern due to the potential for cyberattacks to disrupt space missions, endanger people, and cause international conflicts. With space increasingly becoming a cornerstone of our global infrastructure, implementing robust cybersecurity measures and fostering international cooperation are paramount to safeguarding our collective celestial domain. The future of space must be secure and beneficial for all. Achieving this demands international collaboration, with shared standards and proactive efforts to maintain space as a domain of safety and opportunity [2].

As humanity extends its reach into the cosmos, the domain of space operations has become a critical frontier for scientific discovery, national security, and technological advancement. Space exploration has become vital to our daily lives far beyond exploration, impacting global communications, navigation, and security. However, the expansion into space introduces a unique set of cybersecurity challenges, unlike anything faced on earth. Spacecraft, satellites, and other space-based systems control aspects of space missions from navigation to life support and are potential targets for cyber threats. The realm of space is not just a frontier for exploration but also a battleground for cybersecurity [3].

This chapter provides an introduction to cybersecurity on space vehicles. It begins with giving an overview on cybersecurity. It discusses cybersecurity for space. It presents some applications of cybersecurity in space exploration. It covers space cybersecurity around the world. It highlights some benefits and challenges of cybersecurity in space exploration. The last section concludes with comments.

11.2 OVERVIEW ON CYBERSECURITY

Cybersecurity refers to a set of technologies and practices designed to protect networks and information from damage or unauthorized access. It is vital because governments, companies, and military organizations collect, process, and store a lot of data. As shown in Figure 11.1, cybersecurity involves multiple issues related to people, process, and technology [4]. Figure 11.2 shows different components of cybersecurity [5].

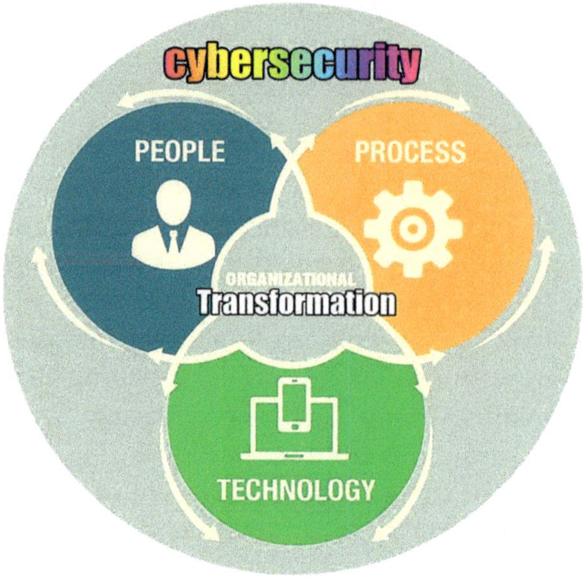

Figure 11.1 Cybersecurity involves multiple issues related to people, process, and technology [4].

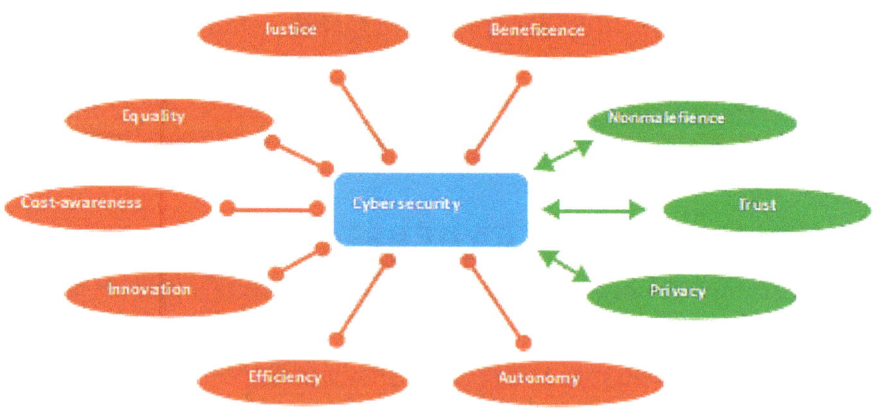

Figure 11.2 Different components of cybersecurity [5].
(Green: supportive; Red: in tension)

A typical cyber attack is an attempt by adversaries or cybercriminals to gain access to and modify their target's computer system or network. Cybercriminals or ethical hackers are modern-day digital warriors, possessing extraordinary skills and knowledge to breach even the most impregnable systems. A typical cybercriminal is shown on Figure 11.3 [6].

Figure 11.3 A typical cybercriminal [6].

Cyber attacks are becoming more frequent, sophisticated, dangerous, and destructive. They are threatening the operation of businesses, banks, companies, and government networks. They vary from illegal crime of individual citizen (hacking) to actions of groups (terrorists) [7].

The cybersecurity is a dynamic, interdisciplinary field involving information systems, computer science, and criminology. The security objectives have been availability, authentication, confidentiality, nonrepudiation, and integrity. A security incident is an act that threatens the confidentiality, integrity, or availability of information assets and systems [8].

- *Availability:* This refers to availability of information and ensuring that authorized parties can access the information when needed. Attacks targeting availability of service generally leads to denial of service.
- *Authenticity:* This ensures that the identity of an individual user or system is the identity claimed. This usually involves using username and password to validate the identity of the user. It may also take the form of what you have such as a driver's license, an RSA token, or a smart card.
- *Integrity:* Data integrity means information is authentic and complete. This assures that data, devices, and processes are free from tampering. Data should be free from injection, deletion, or corruption. When integrity is targeted,

nonrepudiation is also affected.

- *Confidentiality:* Confidentiality ensures that measures are taken to prevent sensitive information from reaching the wrong persons. Data secrecy is important especially for privacy-sensitive data such as user personal information and meter readings.
- *Nonrepudiation:* This is an assurance of the responsibility to an action. The source should not be able to deny having sent a message, while the destination should not deny having received it. This security objective is essential for accountability and liability.

Everybody is at risk for a cyber attack. Cyber attacks vary from illegal crime of individual citizen (hacking) to actions of groups (terrorists). The following are typical examples of cyber attacks or threats [9]:

- *Malware:* This is a malicious software or code that includes traditional computer viruses, computer worms, and Trojan horse programs. Malware can infiltrate your network through the Internet, downloads, attachments, email, social media, and other platforms. Spyware is a type of malware that collects information without the victim's knowledge.
- *Phishing:* Criminals trick victims into handing over their personal information such as online passwords, social security number, and credit card numbers.
- *Denial-of-Service Attacks:* These are designed to make a network resource unavailable to its intended users. These can prevent the user from accessing email, websites, online accounts or other services.
- *Social Engineering Attacks:* A cyber criminal attempts to trick users to disclose sensitive information. A social engineer aims to convince a user through impersonation to disclose secrets such as passwords, card numbers, or social security number.
- *Man-In-the-Middle Attack:* This is a cyber attack where

a malicious attacker secretly inserts him/herself into a conversation between two parties who believe they are directly communicating with each other. A common example of man-in-the-middle attacks is eavesdropping. The goal of such an attack is to steal personal information.

These and other cyber attacks or threats are shown in Figure 11.4 [10]. Sources of cyber threats are displayed in Figure 11.5 [11].

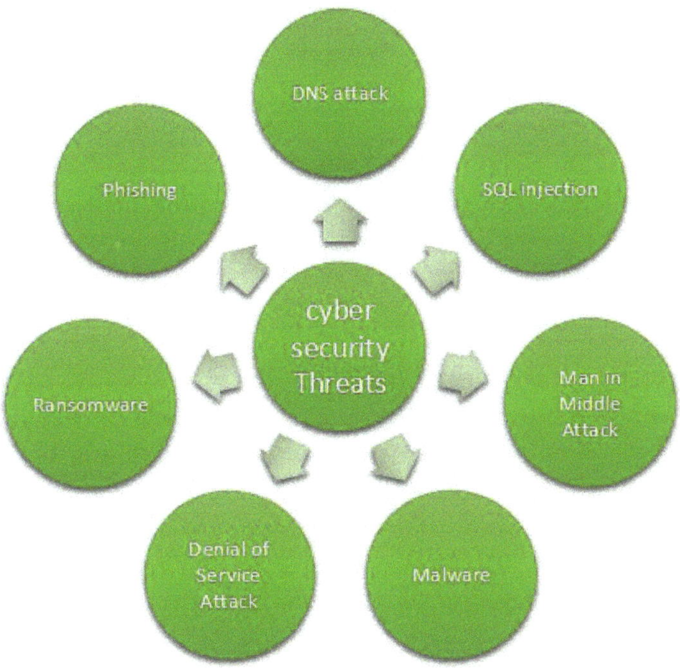

Figure 11.4 Common types of cybersecurity threats [10].

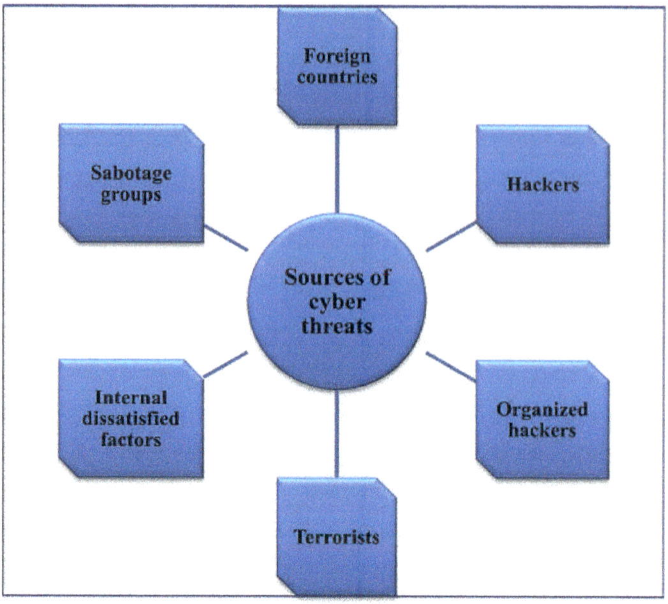

Figure 11.5 Sources of cyber threats [11].

The social and financial importance of cybersecurity is increasingly being recognized by businesses, organizations, and governments. Cybersecurity involves reducing the risk of cyber attacks. Cyber risks should be managed proactively by the management. Cybersecurity technologies such as firewalls are widely available [12]. Cybersecurity is the joint responsibility of all relevant stakeholders including government, business, infrastructure owners, and users. Cybersecurity experts have shown that passwords are highly vulnerable to cyber threats, compromising personal data, credit card records, and even social security numbers. Governments and international organizations play a key role in cybersecurity issues. Securing the cyberspace is of high priority to the US Department of Homeland Security (DHS). Vendors that offer mobile security solutions include Zimperium, MobileIron Skycure, Lookout, and Wandera.

11.3 CYBERSECURITY FOR SPACE

The voyage into space, once the domain of science fiction, has become a testament to human ingenuity and ambition. Space technology has become a key element in communication, navigation, national defense and global surveillance.

Modern society heavily relies on space-based systems for essential functions, making cyberattacks on space infrastructure a significant threat to national security. Nations are depending more and more on space as a mission-critical and developing frontier for information sharing and surveillance. The reliance on space infrastructure brings to the forefront challenges in cybersecurity. Figure 11.6 represents the range of threats to space-based services [13].

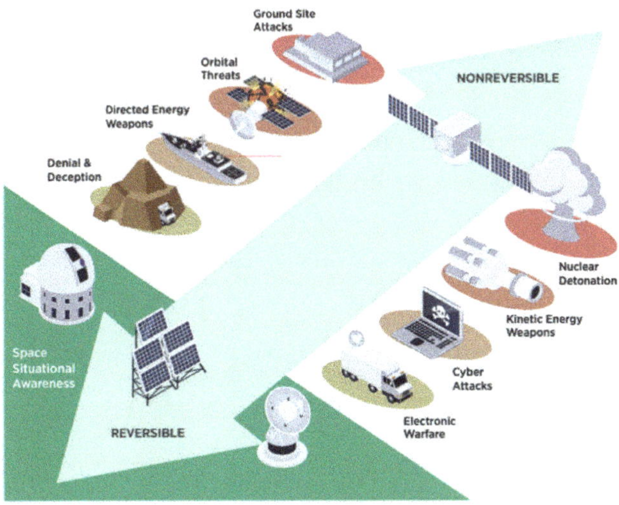

Figure 11 6 The range of threats to space-based services [13].

Cyberattacks on space systems can disrupt global communications, navigation systems, weather forecasting, and military operations, significantly impacting national security and economic stability. Potential cyber threats in space include data manipulation, signal interference, malware installation, ransomware attacks, and disruption of ground control systems. Potential attackers include nation-states, cybercriminals, and even rival companies, each with different motivations for disrupting space operations

Space cybersecurity refers to the growing threat of cyberattacks targeting satellites and other space infrastructure. It presents a new frontier for national security concerns as we rely on space-based systems for communication, navigation, and intelligence increases.

The vulnerabilities of space systems open the door to various cyber attacks. Breaches that threaten communications between ground stations and their space assets are also possible. So is interference with the data streams that flow constantly between satellites and public and private entities. The ramifications of a security breach in such environments range from data loss and compromised missions to the endangerment of human life.

11.4 APPLICATIONS OF SPACE CYBERSECURITY

Outer space has been a national security priority for spacefaring nations since the 1950s.

Space exploration and technology have become integral to modern life, powering critical systems from global communication networks to precision navigation and comprehensive Earth Observation. As space technology becomes more complex and interconnected, the landscape of potential cyberattacks expands. Cyber threats targeting space infrastructure can be broadly viewed as follows [14-16]:

- *Space Systems:* Space missions rely on secure and uninterrupted communication systems. Traditionally, space and terrestrial systems were largely isolated from each other, but this model has changed in recent years. Securing space systems is imperative. Space systems, such as the global positioning system (GPS), support essential services such as military, utilities, aviation, and emergency communications and therefore get drawn into geopolitical conflicts on earth. For example, the Internet increasingly depends on space-enabled communication and information services. Even as the market for space systems evolve, our dependence on space systems for national and economic security will increase dramatically. Space cybersecurity is becoming increasingly important in this new digital era. Protecting the frontier of space systems is unquestionably a security priority for governments and industries. This has increased the number of potential targets for hackers to target, both in space and at control centers located on earth. The national security establishment has recognized that cyber-threats are becoming a greater threat to satellites and communications networks. Keeping space systems secure from malicious attacks is essential in today's world where satellites provide vital services from resilient communications to banking to directing autonomous cars. Figure 11.7 shows a typical space system [13].

Figure 11.7 A typical space system [13].

- *Satellite Interference:* Satellites (from GEO to LEO) are no longer merely amplifying signals but are growing much more complex and playing a much bigger role in communications on earth. They have become indispensable pillars of our modern world that orchestrate navigation, communication, and commerce. Today satellites are pivotal not only in daily conveniences such as the global positioning system (GPS) and Internet access but also in international defense systems and global environmental monitoring. Thousands of satellites are in low Earth orbit, where they are vulnerable to cyberattacks from both above and below. Satellite systems are just as susceptible to cyber attacks as computers on earth. They were not designed with security in mind. The most alarming scenario is attackers gaining complete control of a satellite. They could use it to cause intentional blackouts, trigger international conflicts or even weaponize it against other space assets. Disabling a satellite through cyber means could create space debris that threatens space activities. In 2008, NASA's Terra AM-1 satellite, a key asset for earth observation, experienced two incidents that raised suspicions of cyber interference. The

UN International Telecommunication Union (ITU) prohibits interference with satellite transmissions, yet such interference frequently occurs. Cybersecurity experts, including those at NASA and the European Space Agency, stress the importance of developing robust encryption and secure communication protocols specifically tailored for satellites. Safeguarding our satellites is no longer optional but a necessity for global security and reliability. Satellite communications system is typically displayed in Figure 11.8 [14].

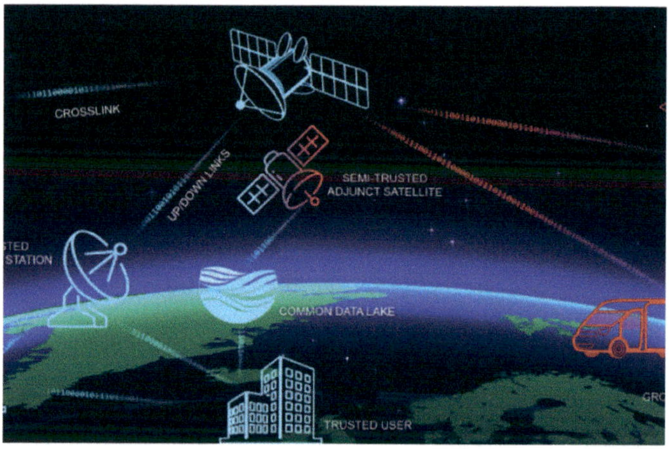

Figure 11.8 Satellite communications system [14.

- *Mega-Constellations:* Mega-constellations (such as SpaceX's Starlink) are vast networks of satellites designed to provide global internet coverage. Their advent represents a significant shift in space technology. Mega-constellations promise enhanced global connectivity, but they also increase the risk of cyber attacks due to a vast attack surface. They pose challenges to protecting space assets from cyber threats. These large constellations also raise environmental concerns, such as light pollution and an increased risk of collisions, adding complexity to their management and regulation. Figure 11.9 illustrates mega-constellation [1].

Figure 11.9 Mega-constellation [1].

- *Space Colonization:* As ambitious plans for space colonization and lunar or Martian habitats gain momentum, the need to secure these extraterrestrial infrastructures becomes more critical. Cybersecurity in these environments goes beyond protecting data; it involves ensuring the operational integrity of life-support systems, habitat controls, and communication links with earth.

- *Global Collaboration:* The complexities of space cybersecurity necessitate a collaborative approach, involving governments, industry, academia, and international bodies. Cybersecurity in space requires global collaboration. The push for international cooperation in using space responsibly is essential. Space technology's future demands robust cybersecurity and international cooperation to safeguard the celestial domain. International collaboration and governance frameworks are essential to building a secure future for space exploration. Collaborations can accelerate the development of space technologies. Successful collaborations, such as one between the University of Alabama in Huntsville and Lockheed Martin, demonstrate how proactive investment in cybersecurity can effectively safeguard satellites. Proactive international collaboration, through initiatives like joint cybersecurity task forces or international space treaties, is essential. These collaborations can foster a secure and sustainable space environment, beneficial for all humanity.

- *Ground-Based Systems:* Space missions rely on ground-based infrastructure for data processing, mission control, and analysis, which can also be vulnerable to cyberattacks. The ground stations are susceptible to both physical and cyber threats. They often use standard communications protocols, making them targets for denial-of-service attacks that can disrupt operations. All ground-based system and network assets should be required to have the following policies in place: incident response, business continuity and crisis communications plans, patching policies, BYOD policies and backup policies. They should be required to join the Space ISAC to be able to collaborate by sharing threats, warnings, and incident information.

11.5 SPACE CYBERSECURITY AROUND THE WORLD

Space is quickly becoming the new frontier to be explored by national governments and private sector actors. Space capabilities are increasing across a growing list of nations, including missile warning, geolocation, and tracking of friendly and adversary activities,

Satellite navigation services—with applications in navigation, munitions guidance, communications, agriculture, banking, and power supply—are critical challenges to security in space to military and civilian users worldwide. At present, most of the economic, commercial, cultural, social and governmental activities and interactions of countries, at all levels, including individuals, non-governmental organizations, government, and governmental institutions, are carried out in cyberspace. Many private companies and government organizations around the world are facing the problem of cyber-attacks and the danger of wireless communication technologies. We can consider how the following nations implement space cybersecurity [17,18]:

- *United States:* National space assets and operations are critical to the security and economic well-being of the United States. The United States has declared that space is now a "war-fighting domain." However, space is not considered a US critical infrastructure in the federal legislature, a list which currently comprises 16 different enterprises. The difficulty of reaching multilateral agreement on cybersecurity and space issues means the United States should address space cybersecurity in plurilateral and bilateral contexts. There are numerous convincing arguments why space needs to be formally listed as US critical infrastructure. The 1991 Gulf War and subsequent U.S. military operations illustrated the value of the US GPS satellite navigation system for troop movement, force tracking, and precision munition delivery. This prompted other nations to develop their own satellite navigation systems.

- *China:* China has also launched multiple ASAT missiles that are able to destroy satellites and developed mobile jammers to deny SATCOM. Beijing's goal is to become a broad-based, fully capable space power. Its rapidly growing space program—second only to the United States in the number of operational satellites—is a source of national pride and part of Chairman Xi Jinping's "China Dream" to establish a powerful and prosperous China. The space program, managed by the People's Liberation Army (PLA), supports both civilian and military interests. China seeks to rapidly achieve these goals through advances in the research and development of space systems and space-related technology. China expects its future wars to be fought mostly outside its borders and in the maritime domain. The PLA uses "informatized warfare" to describe the process of acquiring, transmitting, processing, and using information to conduct joint military operations across the land, sea, air, space, and cyberspace domains. China's space strategy is expected to evolve over time, keeping pace with the application of new space technology. The need for international cooperation has led to the formation of the Asia-Pacific Space Cooperation Organization (APSCO), a multilateral organization with rotating leadership whose members include China, Bangladesh, Iran, Mongolia, Pakistan, Peru, Thailand, and Turkey with Egypt, Indonesia, and Mexico as associate members.

- *Russia:* Russia views space as a war fighting domain and that achieving supremacy in space will be a decisive factor in winning future conflicts. Russia has a strong foundation of technical knowledge and expertise fostered by more than 60 years of experience in space. Moscow's budget is more limited than Beijing's because of competing priorities within Russia's broader military modernization efforts. In the years following the end of the cold war, a combination of economic constraints and technological setbacks caused a decay of Russian space capabilities. However, during the past two decades, Moscow has continued to pursue space services in support of terrestrial applications while developing a suite of counterspace weapon capabilities. Russia is developing a mobile missile that can destroy satellites and crewed space vehicles. Space Strategy and Doctrine Russia openly supports space arms control agreements to prevent weaponization of space. Russian military thinkers believe the importance of space will continue to expand because of the growing role of precision weapons and satellite-supported information networks in all types of conflict. At the same time, Russia is developing an arsenal of counterspace capabilities to attack U.S. and allied assets. As Russia continues to modernize its military, it will increasingly integrate space services into its armed forces. Moscow sees overreliance on space as a potential vulnerability and is determined to avoid becoming excessively dependent on space to conduct its national defense mission. Today the Russian space industry is almost exclusively state owned.

- *Iran:* Iran's pursuit of a national space program supports both its civilian and military goals, including boosting national pride, economic development, and military modernization. Tehran states it has developed sophisticated capabilities, including SLVs as well as communications and remote-sensing satellites. However, its SLVs are only able to launch small satellites into LEO and have proven unreliable. The Iranian Space Agency (ISA) and Iranian Space Research Center (ISRC)—subordinate to the Ministry of Information and Communications Technology— along with the Ministry of Defense and Armed Forces Logistics (MODAFL) oversee part of the country's satellite development programs. ISA and ISRC work with Iranian universities, private industry, and foreign partners to develop satellites to test communications and remote-sensing technologies.

However, Iran's limited space launch capacity has led to a significant backlog of built-but-unlaunched satellites. Iran has improved its domestic space domain awareness capabilities, establishing its first space-monitoring center in 2013. Iran may also contribute to the proliferation of such jamming equipment. Since 2010, state-owned Iran Electronics Industries has marketed several GPS jammers on its website.

- *North Korea:* In 2020, North Korean actors conducted numerous cyberoperations against US foreign partners' defense industries and attempted to compromise various US Government networks. Several North Korean hacker groups have targeted the aerospace industries potentially including space technologies. This activity, if left unchecked, could enable North Korea's weapons and space system development and procurement programs. North Korea's ballistic missiles in theory could be used to target satellites in a conflict. North Korea has placed two satellites in orbit and has articulated further space ambitions. Its space program has also enabled testing of technology used in ballistic missiles under the guise of peaceful use of space.

- *India:* India is developing advanced cybersecurity tools to protect its space systems and software. It is working to collaborate with academia to create a global hub for cybersecurity innovation. Its current effort aims to develop the cybersecurity tools and expertise to protect its space-focused systems and software. On August 23, 2023, the Indian Space Research Organization's (ISRO) performed a soft landing on the moon near the lunar south pole, becoming the fourth country (after the Soviet Union, United States, and China) to perform a soft-landing on the lunar surface and the first to put a spacecraft near the lunar south pole.

11.6 BENEFITS

Cybersecurity for space exploration is important to maintain secure lines of communication and accurate navigation. It underscores the critical need to protect space assets. Safeguarding space assets against cyber threats demands an inclusive strategy that integrates technological advancements, policy frameworks, and public-private partnerships. Cybersecurity for space exploration is also important to maintain secure lines of communication, accurate navigation, and precise control. There will be a greater focus on tools that can enhance space innovation and cyber secureness such as zero trust, software bill of materials, and secure by design. Other benefits of cybersecurity in space include the following:

- *Security:* National security can no longer be defined in terms of military issues and internal and external borders. Today, the risk of declining quality of life of citizens is a threat to national security. The vulnerabilities posed by cyber threats are sporadic, multidimensional. Because they are associated with sensitive networks and infrastructure, their level of damage are very high. Any breach in the security of the networks used for space communications can lead to leaks of sensitive information and/or operational disruption. Cybercriminals can use hacking, jamming, and spoofing to gain control of spacecraft and satellites, steal data, and disrupt communications.
- *Space Reliance:* Modern societies heavily rely on space-based systems for communication, navigation, weather forecasting, intelligence gathering, and military operations, making them crucial for daily life and national security. Space reliance, in an era of competition and expansion, refers to the growing dependence on space-based technologies and services. This occurs simultaneously with a heightened level of competition between nations and companies to access and utilize space. As more players enter the market and develop advanced space capabilities, there are concerns about security and stability in the orbital domain.
- *Space Commercialization:* Private companies are increasingly involved in space activities, including satellite launches, space tourism, and satellite internet, further contributing to the competitive landscape.
- *Encryption:* This is a reversible method of encrypting data that requires a key to decrypt. It provides another level of confidentiality. The system for encrypting and decrypting data is the encryption system. Encryption is a powerful tool for protecting important and private information when exposed to threats from strangers and criminals, as well as for hiding unauthorized activities from law enforcement.

- *Cyber Defense:* This involves the utilization of all unarmed cyber and non-cyber facilities of a country, to create deterrence, prevention, timely detection, effective and deterrent response to any cyber attack.

11.7 CHALLENGES

The exploration of space is fraught with unparalleled challenges, particularly in the realm of cybersecurity. Satellites and other space systems have inherent vulnerabilities that make them prime targets for cyber attacks. Securing space systems presents unique challenges due to the remote location of satellites, limited bandwidth for communication, and the need for robust encryption mechanisms to protect sensitive data. Implementing advanced security measures in space systems is challenging due to limited onboard processing power and bandwidth constraints. Cybersecurity for space has been a challenge due to the older systems being used. Other cybersecurity challenges faced by space exploration include the following [19]:

- *Risks:* Cyberattacks have been a risk for many years and are multifaceted. The complexity and interconnectedness of space assets along with the proliferation of small satellites and commercial ventures heighten the risks involved. As space becomes more important to our critical infrastructure, the impact of a cyber attack and the corresponding risk increases. The risk to commercial space operations needs to be understood and managed alongside other risks to ensure safe and successful operations.
- *Remote Location:* The remote location of space assets, once an advantage, now presents new challenges. Their distance from the earth's surface offers some protection against physical tampering. However, the distance also creates significant challenges for timely intervention and updates. The latency can cause substantial disruptions to critical missions. The reliance on remote operation further exacerbates the vulnerability.
- *Standards:* Developing and adhering to global cybersecurity standards and treaties will be crucial in maintaining the

security and integrity of space missions. To mitigate the risks associated with future space trends, developing shared standards and protocols for space cybersecurity is vital. This includes agreements on the use and protection of space assets and collaborative efforts to track and respond to cyber threats. Collaboration between nations and space agencies is crucial to develop standardized security protocols and share threat intelligence. A defense-in-depth approach that combines policy, standards, and technical solutions is needed to protect space systems.

- *Regulation:* The lack of clear international regulations regarding cyber activity in space creates ambiguity and potential legal complications in addressing space-based cyber threats.
- *Complexity:* The complex nature of space systems and the rapid pace of technological advancements present significant challenges for developing robust cybersecurity measures. As more satellites are launched, the overall space infrastructure becomes more complex and potentially more vulnerable to cyberattacks. The complexity and automation of spacecraft propulsion systems create many entry points for cyberattacks, and compromising these systems could be catastrophic.
- *Data Manipulation:* Altering data transmitted to or from satellites can lead to misinformation or incorrect commands. This could disrupt weather forecasting. Additionally, it could cause navigation systems to malfunction.
- *Signal Jamming:* Disrupting satellites' signals can cripple communication networks, impact military operations, and pose risks to civilian air travel. Jamming could lead to significant economic disruptions and pose risks to public safety. GPS jamming and spoofing are threats to the communication channels that support space operations
- *Spacecraft Vulnerabilities:* One of the biggest space cybersecurity threats is hacking into spacecraft and satellites. Spacecraft are becoming increasingly susceptible to cyberattacks, posing a significant risk to mission success

and astronaut safety. Once launched, it is nearly impossible to physically access a spacecraft for maintenance or updates. This means that cybersecurity risks must be anticipated and mitigated before launch. The space environment exposes spacecraft to extreme conditions, such as radiation, which can affect data integrity and onboard systems.

- *Public Awareness:* Public awareness of the risks and the role of collaborative initiatives in mitigating them are important. It should include supporting policies and organizations dedicated to advancing space cybersecurity and creating education programs for those interested in this subfield. Actions at the national, industry, and international levels can harness growing awareness about space cybersecurity.
- *Potential Impact:* Cyber-attacks on space infrastructure can include malware installation, ransomware attacks, and data or system breaches. They can disrupt communications, supply chains, national security, and more. They can also lead to injury or death of astronauts, property destruction, and halted space exploration progress.
- *Limited Resources:* Without adequate funding and Internet connectivity, a nation may not have the resources to implement adequate security measures. Without the right funding and Internet, most countries around the world are battling to stay ahead in the space race. These limited resources can lead to low or non-existent security measures in place.
- *Insider Threats:* These threats come from individuals with authorized access to space systems and data, and can include human error or malicious intent
- *Cyber Weapon:* A cyber weapon is a system designed and manufactured to damage the structure or operation of other cyber systems. These systems include bot networks, logic bombs, cyber vulnerability exploitation software, malware, and traffic generation systems. Cyber weapons are designed for cyber- warfare Figure 11.10 shows the relationship between cyber warfare, cyber-attack, and cyber-crime [11].

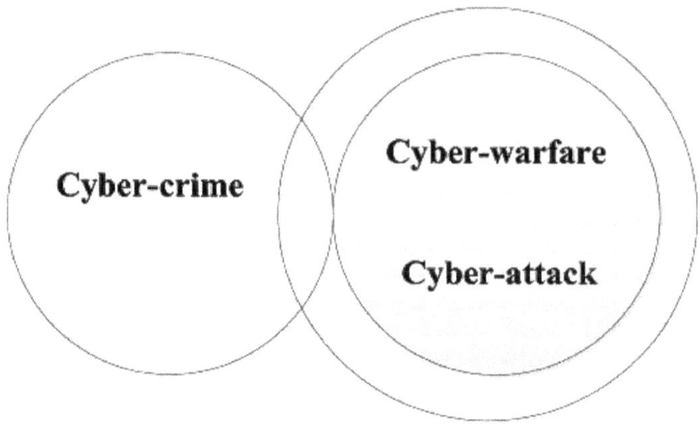

Figure 11.10 The relationship between cyber warfare, cyber-attack, and cyber-crime [11].

11.8 CONCLUSION

Cybersecurity is important for space exploration because space missions are becoming more sophisticated and vulnerable to cyber threats. Cybersecurity must be an integral part of the design and operation of space systems, not an afterthought. The importance of cybersecurity in space exploration cannot be overstated. As we grow increasingly reliant on space technologies, the responsibility to protect them intensifies. Emerging technologies such as AI pose new challenges but also offer opportunities for advancing space cybersecurity.

The journey of space exploration is an ongoing narrative of human achievement and resilience. The future of space exploration promises great discoveries and innovations, but it demands vigilance and commitment to securing this new frontier. It also demands an advanced approach to cybersecurity. More information on cybersecurity in space exploration can be found in the books in [20,21] and a related journal: *Space Policy*.

REFERENCE

[1] Sangfor Technologies, “Space cybersecurity: Exploring challenges, and opportunities,” September 2023,

https://www.sangfor.com/blog/cybersecurity/space-cybersecurity-exploring-challenges-and-opportunities

[2] S. Kaczmarek, “The evolving cybersecurity landscape in space: A new frontier for national security,” April 2024,

https://www.afcea.org/signal-media/cyber-edge/evolving-cybersecurity-landscape-space-new-frontier-national-security

[3] M. N. O. Sadiku, U. C, Chukwu, and J. O. Sadiku, “Cybersecurity in space exploration,” to appear

[4] P. Singh, “A layered approach to cybersecurity: People, processes, and technology- explored & explained,” July 2021,

https://www.linkedin.com/pulse/layered-approach-cybersecurity-people-processes-singh-casp-cisc-ces

[5] M. Loi et al., “Cybersecurity in health – disentangling value tensions,” *Journal of Information, Communication and Ethics in Society*, June 2019,

https://www.emerald.com/insight/content/doi/10.1108/JICES-12-2018-0095/full/html

[6] M. Adams, “Unlocking the benefits of ethical hacking: The importance of ethical hackers in cybersecurity,” April 2023,

https://www.businesstechweekly.com/cybersecurity/network-security/ethical-hacking/

[7] M. N. O. Sadiku, S. Alam, S. M. Musa, and C. M. Akujuobi, “A primer on cybersecurity,” *International Journal of Advances in Scientific Research and Engineering*, vol. 3, no. 8, Sept. 2017, pp. 71-74.

[8] M. N. O. Sadiku, M. Tembely, and S. M. Musa, “Smart grid cybersecurity,” *Journal of Multidisciplinary Engineering Science*

and Technology, vol. 3, no. 9, September 2016, pp.5574-5576.

[9] "FCC Small Biz Cyber Planning Guide,"

https://transition.fcc.gov/cyber/cyberplanner.pdf

[10] "The 8 most common cybersecurity attacks to be aware of,"

https://edafio.com/blog/the-8-most-common-cybersecurity-attacks-to-be-aware-of/

[11] Y. Li and Q. Liu, "A comprehensive review study of cyber-attacks and cyber security; Emerging trends and recent developments," *Energy Reports*, vol. 7, November 2021,

https://www.sciencedirect.com/science/article/pii/S2352484721007289

[12] Y. Zhang, "Cybersecurity and reliability of electric power grids in an interdependent cyber-physical environment," *Doctoral Dissertation*, University of Toledo, 2015.

[13] "2022 Challenges to security in space: Space reliance in an era of competition and expansion,"

https://www.dia.mil/Portals/110/Documents/News/Military_Power_Publications/Challenges_Security_Space_2022.pdf

[14] S. Kaczmarek, "We need cybersecurity in space to protect satellites," February 2024,

https://www.scientificamerican.com/article/we-need-cybersecurity-in-space-to-protect-satellites/

[15] "Cybersecurity protections for spacecraft: A threat based approach,"

https://aerospace.org/paper/cybersecurity-protections-spacecraft-threat-based-approach

[16] C. Brooks, "Space, satellites and cybersecurity," March 2022,

https://www.linkedin.com/pulse/space-satellites-cybersecurity-chuck-brooks

[17] C. Brooks, "Cybersecurity of space systems," February 2024,

https://www.linkedin.com/pulse/cybersecurity-space-systems-chuck-brooks-i0c3e

[18] R. Lemos, "India needs better cybersecurity for space, critical infrastructure," September 2024,

https://www.darkreading.com/ics-ot-security/india-cybersecurity-space-critical-infrastructure

[19] S. Kaczmarek, "Cybersecurity challenges in space exploration," February 2024,

https://www.cutter.com/article/cybersecurity-challenges-space-exploration

[20] M. N. O. Sadiku, *Cybersecurity and Its Applications*. Moldova, Europe: Lambert Academic Publishing, 2023.

[21] J. G. Oakley, *Cybersecurity for Space: Protecting the Final Frontier*. Apress, 2020.

CHAPTER 12
GAMIFICATION IN SPACE EXPLORATION

"Gamification is the craft of deriving fun and engaging elements found typically in games and thoughtfully applying them to real-world or productive activities."

—Yu-kai Chou

12.1 INTRODUCTION

Games are widely popular and entertaining. The games industry is one of the most lucrative industries due to the billion dollar sales of digital games. The global game marketplace includes video game console hardware and software and online, mobile, and PC games. Games are designed systematically, thoughtfully, and artistically for the purpose of creating fun and enjoyment. Figure 12.1 shows various uses of games.

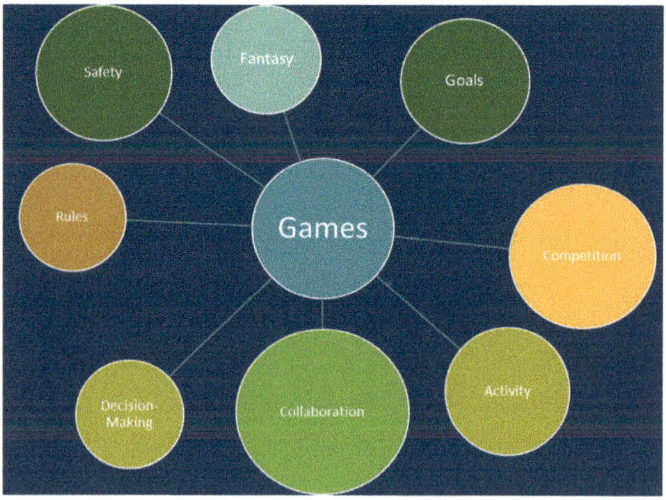

Figure 12.1 Different uses of games.

The difference between regular and "serious" games is that the latter promote "serious" purposes, such as learning and motivation.

Although games and gamification have a lot in common, they are not exactly the same Gamification, as opposed to serious gaming, is the process of applying the science and psychology of digital gaming (such video game elements) in a non-game environment. It is the craft of deriving all the fun in games and applying them to productive activities [1].

Gamification is a strategy that uses game-like elements to make learning more engaging and interactive. Gamification in space exploration represents a relatively recent but innovative approach to astronaut training and public engagement. It has been used in space education to help students learn about space and develop skills for careers in the space sector.

Gamification is a technique that incorporates game-like elements into non-game contexts to make learning more engaging. It is used to engage the public, especially students, in space missions. It can be used to train astronauts and educate the public about space. Gamification elements, such as competition, rewards, and interactive challenges, can be used to encourage exploration. Space exploration in games can be mind-numbing or jaw-dropping, depending on one's expectations [2,3].

The aim of this chapter is to describe how the gamification approach has been applied within the space sector. It begins with describing what gamification is all about. It covers gamification in space exploration. It presents some applications of gamification in space exploration. It highlights some benefits and challenges of gamification in space exploration. It discusses space gamification around the world. The last section concludes with comments.

12.2 WHAT IS GAMIFICATION?

The word "gamification" was coined in 2002 by Nick Pelling, a British inventor, but it did not gain popularity until 2010. The idea of gamification came from the fact that the gaming industry was the first to master human-focused design and we are now learning from games. Gamification is not a new concept, but it is deeply rooted in marketing endeavors, such as points cards, grades, and degrees, and workplace productivity [4]. Researchers became interested in gamification because the concept could be implemented in different ways to motivate people. Gamification has become hugely popular in all walks of life, including education. The concept of gamification is illustrated in Figure 12.2 [5].

Figure 12.2 The concept of gamification [5].

The gamification can be viewed in two ways: (1) adopting the act of playing a video game into everyday use, (2) the act of using game elements to make non-games more enjoyable. It is applied in education, business, sports, marketing, and finance. It is currently one of the largest trends in education. Traditional education has been found to be ineffective in motivating and engaging many students. Gamification is cutting-edge approach which is producing positive results in every region of the world.

Everyone loves games. Gamification just takes advantage of that innate desire. It can make practice fun. It can make the routine less droll. Online games have become bolder and more diverse. Gamification involves the introduction of gameplay to a traditionally non-game environment. Its goal is to encourage user engagement. There are several organizations that use gamification to engage users. These include US Army. Nile, Hilton Garden, McDonald, and Starbucks. The impact of gamification on employees is demonstrated in Figure 12.3 [6].

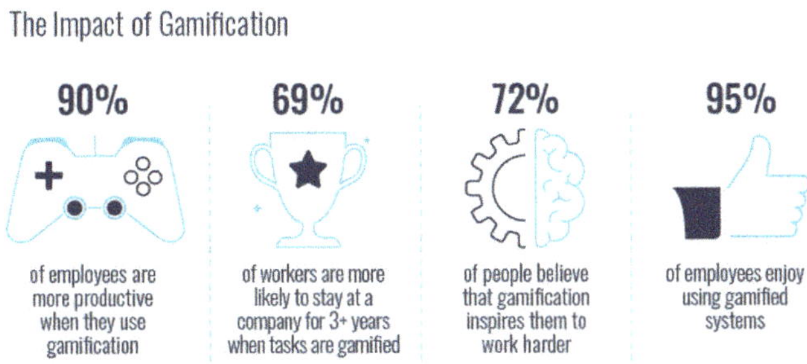

Figure 12.3 The impact of gamification on employees [6].

12.3 GAMIFICATON IN SPACE EXPLORATION

The cosmos has long been an irresistible frontier, beckoning humanity to explore its vast mysteries. Space is the final frontier, and it is not an easy or cheap to explore. It is full of mystery, as typically shown in Figure 12.4 [7].

Figure 12.4 Space is full of mystery [7].

We, as humans, are driven to try to solve all the mysteries that we can. But we must prepare and train astronauts before we can embark on the interstellar odyssey. The preparation for space travel is not only demanding but also deeply complex, involving a myriad of skills. This is where gamification comes in. Gamification has recently emerged as a creative and powerful tool for engaging learners and improving the effectiveness of education. It is a concept that involves applying game-like elements, such as competition, rewards, and interactive challenges, to non-game contexts to engage and motivate individuals. Some space companies are taking the lead, harnessing the power of gamification to transform astronaut training, Figure 12.5 shows one of the most popular space exploration games, known as No Man's Sky [8].

Figure 12.5 A popular space exploration game, known as No Man's Sky [8].

Some examples of gamification in space exploration include [9]:

- *Space Simulations:* The use of space simulations, which replicate the conditions and challenges of space travel, predates formal gamification efforts. Early astronaut training programs often incorporated simulation-based training exercises.
- *Apollo Exploration Missions:* A serious game that simulates lunar roving activities, such as planning routes, loading equipment, and driving the rover. The game uses 3D terrains and spacecraft models to create a realistic environment.
- *NASA Space Voyagers:* A game where players take turns using spacecraft and instruments to explore a destination. Players tally research points, roll dice to get challenges, and take actions based on the destination card. Figure 12.6 illustrates the Voyager game [10].

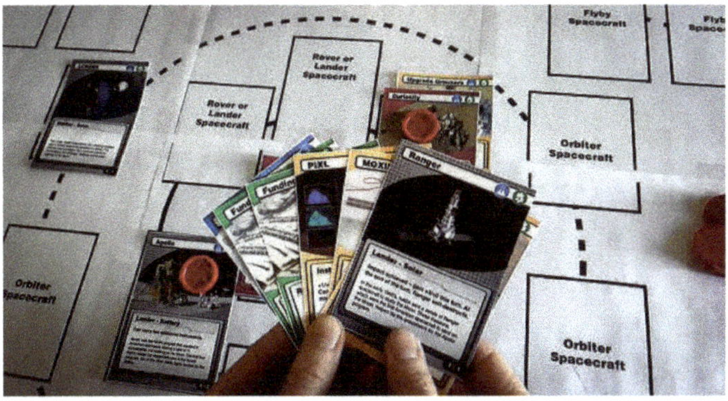

Figure 12.6 NASA Voyager game [11].

- *MapUncover:* A mobile map app that uses gamification to encourage users to explore. The app has a leaderboard that limits the number of visible users to 100, and the top three users receive a trophy.

12.4 APPLICATIONS OF GAMIFICATON IN SPACE EXPLORATION

Game is a fundamental human practice that can spark curiosity and facilitate skills acquisition and behavioral change. Gamification (or gameful design) is a popular motivation enhancement method. It is regarded as the use of game elements and game-design techniques in non-game contexts. It operates in space exploration in the following ways [9,11]:

- *Astronaut Training:* One area that has seen significant interest in gamification is the training of the astronauts. Space agencies and training centers develop realistic simulations of space missions, including microgravity environments and extravehicular activities. These simulations often include game-like challenges, such as repairing equipment or conducting experiments. Gamified training systems track astronauts' progress, awarding points or achievements for completing tasks or demonstrating proficiency in specific skills. This tracking helps astronauts monitor their development and identify areas for improvement. Astronauts may participate in friendly competitions, either individually or as part of a team. Competition fosters a sense of achievement and encourages continuous improvement.

- *Space Education:* Gamification, a modern learning approach, has garnered attention for its potential to spark space science education. The gamification approach empowers educators to build stronger collaboration and interaction among students and teachers. Integrating gamification into the educational landscape can revolutionize traditional learning paradigms. Space education's gamification can help inspire students to learn about space and build the knowledge, vision, and skills. Space agencies and educational institutions create space-themed games and apps for the public. These games teach astronomy, physics, and space science engagingly and entertainingly. Space agencies run online challenges and competitions that allow the public to tackle real-world space problems. Participants may be asked to propose solutions to specific challenges, with rewards or recognition for successful ideas. Gamification can also be used to make learning more interactive and engaging by breaking learning down into smaller, achievable tasks. Common gamification elements include points, timers, badges, and leaderboards. Figure 12.7 shows aeronautic practice-space for engaging students in learning activities [12].

Figure 12.7 Aeronautic practice-space for engaging students [12].

- *Problem-Solving:* Private space companies and organizations often host innovation challenges with substantial cash prizes. These challenges encourage participants to develop innovative solutions to complex problems in space exploration, employing gamification to motivate inventors. Gamification can be used to crowdsource ideas and solutions for space-related issues. Challenges and competitions provide incentives for participants to contribute their expertise and creativity. Figure 12.8 shows the workflow of the space exploration learning experience [11].

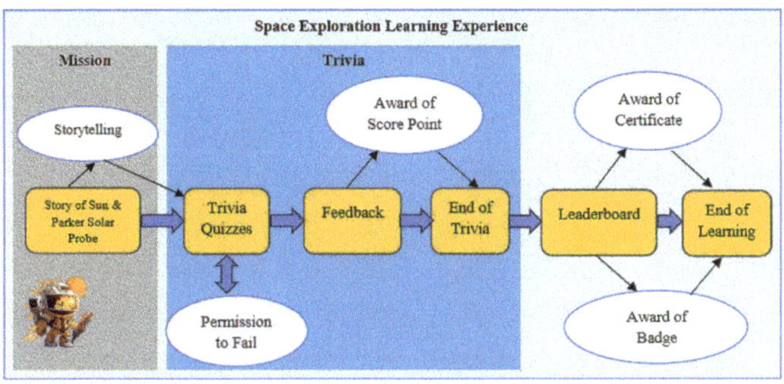

Figure 12.8 Workflow of the space exploration learning experience [11].

- *Virtual Reality Training:* While not traditionally gamification, virtual reality is used to create immersive, game-like training simulations. Astronauts can practice critical tasks in a realistic but controlled environment, enhancing their readiness for space missions. Advanced simulations, perhaps incorporating virtual reality (VR) and augmented reality (AR), will provide immersive and highly realistic training experiences.

12.5 BENEFITS

Gaming is not just for kids, so to speak. Using gamification in and of itself is training students for how they may conduct tasks in their professional careers. Gamification whether it uses space as its arena or not, can be a valuable guidance tool for educators looking to implement gamification in their teaching. People and agencies in the space sector have taken an interest in the gamification. Gamification can help with the mental and physical well-being of astronauts in space. Astronauts can experience loneliness and boredom in microgravity environments, so games and recreational activities can help with enjoyment and relaxation. Other benefits of space gamification include the following [9,13,14]:

- *Advanced Training:* Gamification will play an even more critical role in astronaut training. Advanced simulations, perhaps incorporating virtual reality (VR) and augmented reality (AR), will provide immersive and highly realistic training experiences. These simulations will enable astronauts to practice complex tasks and emergencies, enhancing their preparedness for space missions.

- *Space Education:* Generally in education, gamification is used to increase the engagement and motivate students to learn. Gamification will continue to be a powerful tool for engaging the public in space exploration. More interactive educational games and citizen science projects will emerge, allowing individuals to actively contribute to scientific research and deepen their understanding of space. One of the main benefits of gamification in space education is that it can help to motivate learners by using an exciting and thought-provoking endeavor. This can help to motivate students to invest more time and effort into learning about space. Another benefit of gamification in space education is that it can help to make learning more interactive.

- *Passion for STEM:* Gamification approach is used within the space sector to develop online and offline resources to foster passion and interest in STEM (Science, Technology, Engineering and Mathematics)-related subjects. Indeed, a decreasing number of students nowadays decide to take up STEM-related studies and careers. Therefore, following an edutainment logic, it is essential to design more gamification activities to increase students' motivation and engagement with STEM topics.

- *Astronaut Training:* Space agencies and training centers develop realistic simulations of space missions, including microgravity environments and extravehicular activities. These simulations often include game-like challenges, such as repairing equipment or conducting experiments. Gamified training systems track astronauts' progress, awarding points or achievements for completing tasks or demonstrating proficiency in specific skills.

This tracking helps astronauts monitor their development and identify areas for improvement.

- *Problem Solving:* Private space companies and organizations often host innovation challenges with substantial cash prizes. These challenges encourage participants to develop innovative solutions to complex problems in space exploration, employing gamification to motivate inventors. Gamification can also be used to crowdsource ideas and solutions for space-related issues. Challenges and competitions provide incentives for participants to contribute their expertise and creativity.

- *Remote Space Missions:* As space agencies plan for more extended missions to distant destinations like Mars, gamification will become essential for maintaining crew morale, mental health, and engagement during extended periods of isolation and confinement. Gamified activities and challenges can help combat boredom and stress.

- *Virtual Space Tourism:* As commercial space tourism grows, gamification will be integrated into the passenger experience. Passengers may engage in simulated missions or space-themed games during their journeys to make the experience more exciting and informative.

- *International Collaboration:* Gamification can serve as a bridge for international collaboration in space exploration. Jointly developed gamified projects and competitions can foster cooperation among space agencies and organizations worldwide.

- *Lunar Exploration:* Games can be combine with information about the Moon, scientific principles, realism, and intelligent applications of technology to offer an engaging interaction and play. It challenges users to accomplish a lunar exploration mission, which imitates what astronauts did with the Apollo Lunar Roving Vehicle (LRV). The gameplay mechanics is developed based on the actual activities performed by the astronauts on the lunar surface. Visualization of the Earth, Moon, or solar system has been used in many space themed games.

• *Brand Loyalty:* Gamification has been found to be positively associated with brand loyalty, brand attitude, brand awareness, brand engagement, brand involvement, and brand love.

12.6 CHALLENGES

The transformative potential of gamification in space education is not without its challenges. Space is a unique environment! It is so scary on its own because it is so big, so dark, and so filled with unknown things that we cannot even start to comprehend. As humanity ventures further into space, leveraging the unique conditions of microgravity becomes increasingly imperative. Other challenges of space gamification include the following [15]:

• *Digital Divide:* This denotes a division between people who have access and use digital media (e.g., the Internet, mobile, PC, etc.) and those who do not. Typically, this gap is more prominent in developing nations where the socioeconomic living conditions are worse. The digital divide which may become a limit in designing online gamification activities in developing nations.

• *Risk:* Without the extensive knowledge in space travel, it may be difficult for a civilian to be fully informed of the risks of a trip to space, even with the FAA's required Q&A session. Gamification risks to over-emphasize rewards may inadvertently erode intrinsic motivation.

• *Regulation:* Regulation of preparation is also an issue when sending a civilian to space, especially if that preparation is part of a televised competition. From an international law perspective, the outdatedness of widely adopted treaties regarding outer space results in the complete exclusion of any provisions mentioning the relatively new concept of civilian space travel or space tourism.

• *Complexity:* Player engagement is a multifaceted construct that goes beyond mere participation, encompassing emotional, cognitive, and behavioral dimensions. Designing gamification elements that resonate with a diverse learner profile is a significant challenge. A one-size-fits-all approach may lead to disengagement among learners who do not find the gamified elements motivating.

The successful adoption and advancement of gamification in educational settings hinge significantly on the perspectives of its primary end-users.

- *Gamification in Schools:* Teaching space science in schools can be challenging, while on the other hand, students feel a special attraction towards hands-on activities. The gamification approach empowers educators to build stronger collaboration and interaction among students and teachers. By integrating gamified elements and fostering awareness of the solar system, educators can inspire students, elevate their understanding of space, and trigger curiosity about the mysteries of the universe.

12.7 SPACE GAMIFICATION AROUND THE WORLD

Gamification is approaching maturity in the education and enterprise retail sectors. It is being adopted across many industries, including education, healthcare, marketing, and employee training. It will continue to grow in popularity across a variety of industries. We will also continue to see gamification's impact in the space sector. Africa shall see the fastest growth rate, with gamification in the region forecasted to expand by 60.1% by 2023. We consider how gamification is applied in the space exploration of the following nations [6,13]:

- *United States:* National Aeronautics and Space Administration (NASA) is an independent agency of the US federal government responsible for the civilian space program, as well as aeronautics and space research. NASA has used video games to train astronauts to handle medical issues in space. Large enterprises are expected to hold the largest market share in gamification. Companies such as Oracle, Hewlett Packard, IBM, and Google will help drive this trend.

- *Europe:* In Europe, the European Space Agency (ESA) project helps to foster interest and passion for STEM-related studies and to stimulate young people's awareness of Europe's space program. The ESA Education Office has launched the ESA Kids (2021) website. ESA Kids targets primary school kids from ages 3 to 10 years old. This e-learning platform offers many didactic resources presented through a gamified setup. Such projects demonstrate how gamification can make space education more engaging and interactive and inspire learners to explore the world of space. In Europe, the support of local governments and key stakeholders will help to bolster this growth. For example, policymakers in the United Kingdom and France will host more gamification conferences and events.

- *China:* China has made great strides in exploring space in recent years, rocketing astronauts to its own space station and bringing back rocks from the moon. Now it wants to turn those feats into scientific advances. Museums, which are among the most popular science institutions outside schools, usually display and introduce historical culture and cultural relics to tourists
- *India:* As India is now reaching Moon, Mars, and beyond, with that it is extremely important for the young minds of the 21st Century to learn about space science and its possible applications. The Indian Space Research Organization (ISRO) aims at enhancing space education and awareness among students and the general public.

12.8 CONCLUSION

Space exploration carried out by astronauts has inspired younger generations to study science and engineering. Everyone loves to know about the rockets and space travel. To rekindle the youth interest in STEM and prepare them for the growth of STEM occupations, serious games have been made by science museums and space centers. The integration of gamification techniques into various sectors, such as exploration sector, has been a notable trend. Gamification will play a critical role in astronaut training and public education. It has become a popular form to promote learning engagement. The future of gamification in space exploration holds significant promise, as it continues to evolve and expand its influence in various aspects of the field. More information on the integration of gamification into space exploration is available from the books in [16,17] and the following related journals:

- *Space Policy*
- *Journal on Emerging Technologies*

REFERENCE

[1] M. N. O. Sadiku, S.M. Musa, and R. Nelatury, ”Digital games,” *International Journal of Research and Allied Sciences*, vol. 1, no. 10, Dec. 2016, pp. 1,2.

[2] M. N. O. Sadiku, P. A. Adekunte, and J. O. Sadiku, “Gamification in the space exploration,” *International Journal of Trend in Scientific Research and Development*, vol. 8, no. 6, November-December 2024, pp. 1-8.

[3] M. N. O. Sadiku, S. R. Nelatury, and S. M. Musa, “Gamification: A primer,” *Journal of Scientific and Engineering Research*, vol. 7, no. 4, 2020, pp. 165-169.

[4] K. Seaborn and D. I. Fels, “Gamification in theory and action: A survey,” *International Journal of Human-Computer Studies*, vol. 74, 2015, pp. 14–31.

[5] E. Hagene, “Gamification in training: 4 ways to bring games into the classroom,” June 2019,

https://frontcore.com/blog/gamification-in-training-4-ways-to-bring-games-into-the-classroom/

[6] H. Cloke, “19 Gamification Trends for 2023-2025: Top Stats, Facts & Examples,” May 2023,

https://www.growthengineering.co.uk/19-gamification-trends-for-2022-2025-top-stats-facts-examples/

[7] W. Quick, “To boldly go: best space exploration games,” July 2024,

https://www.thegamer.com/best-space-exploration-games/

[8] “Nine most popular space exploration games you must try in 2023,” February 2023,

https://spaceimpulse.com/2023/02/28/space-exploration-games/

[9] “Gamification in space exploration: How dedicated companies turn astronaut training into a game,”

https://smartico.ai/gamification-in-space-exploration/

[10] NASA Space Voyagers: The game,"

https://www.jpl.nasa.gov/edu/teach/activity/nasa-space-voyagers-the-game/

[11] P. C. Chiu et al., "The effects of gamification in space exploration learning," *Journal of Advanced Research in Applied Sciences and Engineering Technology*, vol. 59, no. 1, 2026, pp. 265-274.

[12] "Scientix and Space part I. – Hands-on activities related to space exploration in pre-school and secondary school," April 2028.

https://blog.scientix.eu/2018/04/scientix-and-space-part-i-hands-on-activities-related-to-space-exploration-in-pre-school-and-secondary-school/

[13] A. Galeone, "Gamification in the space sector: How gamification activities can foster the passion and interest for STEM-related studies," 2022

https://www.igi-global.com/chapter/gamification-in-the-space-sector/308758

[14] "Gamification in space exploration: How dedicated companies turn astronaut training into a game,"

https://smartico.ai/gamification-in-space-exploration/

[15] A. Christopoulos and S. Mystakidis, "Gamification in education,"

https://www.mdpi.com/2673-8392/3/4/89

[16] M. N. O. Sadiku, *Gamification and Its Applications*. Moldova, Europe: Lambert Academic Publishing, 2023.

[17] V. A. Oojorah and W. Udhin, *Imagining the Futures of Higher Education in Southern Africa: An Exploration from Multiple Perspectives*. Springer, 2024.

APPENDIX[1] - LIST OF ACRONYMS

AI	ARTIFICIAL INTELLIGENCE
AR	AUGMENTED REALITY
AS	AUTONOMOUS SYSTEMS
DHS	DEPARTMENT OF HOMELAND SECURITY
DTN	DELAY-TOLERANT NETWORKING
ES	EARTH STATIONS
ESA	EUROPEAN SPACE AGENCY
FAA	FEDERAL AVIATION ADMINISTRATION
GEO	GEOSTATIONARY EARTH ORBIT
GPS	GLOBAL POSITIONING SYSTEM
GPU	GRAPHICS PROCESSING UNITS
GSI	GROUND SATELLITE LINKS
HAP	HIGH-ALTITUDE PLATFORMS
HPC	HIGH PERFORMANCE COMPUTING
ICN	INFORMATION CENTRIC NETWORKING

1 Mostly taken from J. Kua et al., "Internet of things in space: A review of opportunities and challenges from satellite-aided computing to digitally-enhanced space living," *Sensors*, vol. 21, no. 23, December 2021.

IEEE	INSTITUTE OF ELECTRICAL AND ELECTRONICS ENGINEERS
IETF	INTERNET ENGINEERING TASK FORCE
IIOT	INDUSTRIAL INTERNET OF THINGS
IOST	INTERNET OF SPACE THINGS
IOT	INTERNET OF THINGS
IP	INTERNET PROTOCOL
ISS	INTERNATIONAL SPACE STATION
ISL	INTER-SATELLITE LINKS
LAN	LOCAL AREA NETWORK
LAP	LOW ALTITUDE PLATFORMS
LEO	LOW-EARTH ORBIT
LORA	LONG-RANGE
LPWAN	LOW-POWER WIDE-AREA NETWORK
M2M	MACHINE-TO-MACHINE
MAC	MEDIUM ACCESS CONTROL
MAP	MID-ALTITUDE PLATFORMS
MEC	MOBILE/MULTI-ACCESS EDGE COMPUTING
MEO	MEDIUM-EARTH ORBIT
ML	MACHINE LEARNING
MIMO	MULTIPLE INPUT MULTIPLE OUTPUT
MMWAVE	MILLIMETRE-WAVE

MTC	MACHINE TYPE COMMUNICATIONS
MQTT	MESSAGE QUEUING TELEMETRY TRANSPORT
NASA	NATIONAL AERONAUTICS AND SPACE ADMINISTRATION
NB-IOT	NARROWBAND-IOT
NDN	NAMED DATA NETWORKING
NFV	NETWORK FUNCTION VIRTUALIZATION
NNI	NATIONAL NANOTECHNOLOGY INITIATIVE
OMNI	OPERATING MISSIONS AS NODES ON THE INTERNET
PLA	PEOPLE'S LIBERATION ARMY
PTO	PROXIMAL TERRESTRIAL OFFLOADING
PULSAR	PROTOTYPE OF AN ULTRA LARGE STRUCTURE ASSEMBLY ROBOT
QOS	QUALITY OF SERVICE
QUIC	QUICK UDP INTERNET CONNECTIONS
RFID	RADIO-FREQUENCY IDENTIFICATION
RTO	REMOTE TERRESTRIAL OFFLOADING
SATCOM	SATELLITE COMMUNICATIONS

SBO	SATELLITE-BORNE OFFLOADING
SCAN	SPACE COMMUNICATIONS AND NAVIGATION
SAGIN	SPACE, AIR, GROUND INTEGRATED NETWORK
SDN	SOFTWARE DEFINED NETWORKING
SN	SPACE NETWORK
TCP	TRANSMISSION CONTROL PROTOCOL
TN	TERRESTRIAL NETWORKS
UAV	UNMANNED AERIAL VEHICLES
UDP	USER DATAGRAM PROTOCOL
WAN	WIDE AREA NETWORK

INDEX

www.ingramcontent.com/pod-product-compliance
Lightning Source LLC
LaVergne TN
LVHW052109210926
848482LV00024B/386